Scientific Technology & Modern Society

Dr. Vera Pavri
Ryerson University

Kendall Hunt
publishing company

Cover image © Shutterstock, Inc.

Kendall Hunt
publishing company

www.kendallhunt.com
Send all inquiries to:
4050 Westmark Drive
Dubuque, IA 52004-1840

Copyright © 2010 by Kendall Hunt Publishing Company

ISBN 978-0-7575-8273-8

Printed in Canada
10 9 8 7 6 5 4 3

Table of Contents

Introduction

Have you ever fallen asleep reading an article given to you in class? Have you ever found yourself reading the same line over and over again and still do not understand what you have read? This is a common problem most students have when tackling articles and book chapters. One of the reasons for this is because students try to read materials assigned to them much in the same way they try to read books or magazines they choose for themselves for pleasure. THIS DOES NOT WORK. In order to keep your mind focused and active on an article that you would not necessarily choose to read yourself, you must be aware of why you have been assigned that article and what approach you are going to take when you finally decide to try and read it.

When reading any scholarly article, the important thing to remember is PURPOSE. Why are you reading the article in the first place? Is it to gain new knowledge or supplement ideas that have already been introduced to you in lecture? In this class, unless stated otherwise, all the readings are designed to supplement knowledge – that is, to further ingrain information that has already been presented in class. As such, a good approach to reading in this class is to tackle articles following a lecture. After all, how will you know what to look for in a reading if you don't know what are the most important concepts that have been covered in lecture?

If you attend the lecture and are comfortable with the material that has been introduced to you, you are probably going to be reading to supplement knowledge. Here, you are actively looking for material that is familiar to you and which can be used to solidify ideas and arguments in your mind. This way, you are doing more than just memorizing facts, which will soon be forgotten – you are actually learning (and hopefully retaining) important information. This will allow you to practice the art of skimming texts, which takes both time and practice to perfect.

In contrast, if you missed the lecture or really didn't understand what was being said, then you will be reading to gain new knowledge. It will take you longer to get through an article. This is because you are being introduced to new concepts and ideas that are unfamiliar to you, and this simply requires more time and effort.

During this year, I will be providing you the opportunity to learn ways to better your critical reading skills. Our first tutorial seminar will provide you with fundamental information of what to look for when reading any academic work, such as key words or important sentences. Following this tutorial, other examples will also be provided in lecture about critical reading skills, so that hopefully by the end of term, you won't be falling asleep as frequently as you have in the past!

As for the readings themselves, here are a few things to consider. Remember that all of our discussions in this class will focus on one or more of the course theories and themes that were presented to you in the first lecture. Articles such as Sergio Sismondo's "Two Questions Concerning Technology" and Sungook Hong's "Historiographical Layers in the Relationship between Science and Technology" are especially important when trying to understand theories and themes in our course such as technological determinism, technology as applied science, and the relationship between science and technology, while Nelly Oudshoorn and Trevor Pinch's "How Users and Non-users Matter," Claude Fischer's "'Touch Someone': The Telephone Industry Discovers Sociability," Paul Ceruzzi's "The Personal Computer, 1972–1977," and Jan van den Ende et al.'s "Shaping the Early Development of Television" are excellent accounts of looking at how users shape new technology.

Lynne Osman Elkin's article "Rosalind Franklin and the Double Helix" provides us with a vivid illustration of gender and science issues and Nathan Rosenberg's "Why in America" offers a study of science and technology from both a historical and a cross-culture perspective. Robert Proctor's article "Nazi Science and Nazi Medical Ethics: Some Myths and Misconceptions" is an excellent piece that tackles the issue of science, technology, and race, while the role of politics in science and technology decisions is made clear in articles such as Barton Bernstein's "The Atomic Bombings Reconsidered" and Michael Gordin et al.'s "Ideologically Correct Science." Finally, articles like David Hounshell's "Mass Production," Thomas Hughes "The System Must Be First," Lawrence Lessig's "Code Is Law," the U.S. Department of Energy's "Genomics and Its Impact on Society," and Peter Bowler and Iwan Rhys Morus' "Ecology and Environmentalism" all discuss the theme of management and control – whether it be machines, natural resources, or the human body.

Good luck and happy reading!

Dr. Vera Pavri

Section 1

Science, Technology and the Birth of Modern Industry

Two Questions Concerning Technology

Reading 1

Sergio Sismondo

■ Is Technology Applied Science?

The idea that technology is applied science is now centuries old. In the early seventeenth century, Francis Bacon and René Descartes both argued for the value of scientific research by claiming that it would produce useful technology. In the twentieth century this view was championed most importantly by Vannevar Bush, one of the architects of the science policy pursued by the United States after the Second World War: "Basic research . . . creates the fund from which the practical applications of knowledge must be drawn. New products and new processes do not appear full-grown. They are founded on new principles and new conceptions, which in turn are painstakingly developed by research in the purest realms of science. . . . Today, it is truer than ever that basic research is the pacemaker of technological progress."

The view that technology is applied science has been challenged from many directions. In particular, accounts of artifacts and technologies show that scientific knowledge plays relatively little direct role in the development even of many state-of-the-art technologies. Historians and other theorists of technology have argued that there are technological knowledge traditions that are independent of scientific knowledge traditions, and that to understand the artifacts one needs to understand those knowledge traditions. At the same time, however, some people working in science and technology studies (S&TS) have argued that science and technology are not sufficiently well defined and distinct for there to be any determinate relationship between them.

Because of its large investment in basic research, in the mid-1960s the US Department of Defense conducted audits to discover how valuable that research was. Project Hindsight was a study of the key events leading to the development of 20 weapons systems. It classified 91 percent of the key events as technological, 8.7 percent as applied science, and 0.3 percent as basic science. Project Hindsight thus suggested that the direct influence of science on technology was very small, even within an institution that invested heavily in science, and which was at some key forefronts of technological development. A subsequent study, TRACES, challenged that picture by looking at prominent civilian technologies and following their origins further back in the historical record.

Among historians of technology it is widely accepted that "science owes more to the steam engine than the steam engine owes to science." For example, work on the history of aircraft suggests that aeronautical engineering is relatively divorced from science: engineers consult scientific results when they see a need to, but there is no sense in which their work is driven by science or in which it is the application of science (Vincenti 1990). Engineers develop their own mathematics, their own experimental results, and their own techniques. Or, similarly, the innovative electrical engineer Charles Steinmetz did not either apply physical theory or derive his own theoretical claims from it (Kline 1992), instead developing theoretical knowledge in purely engineering contexts. This is despite the fact that he considered his work to be applied science; it was applied science in that it fit Steinmetz's understanding of the scientific method.

Edwin Layton (1971, 1974) argues that the reason why the model of technology as applied science is so pervasive, and yet individual technologies cannot be seen to depend on basic science, is that technological knowledge is downplayed. Engineers and inventors participate in knowledge traditions, which shape the work that they do. Science, then, does not have a monopoly on technical knowledge. In the nineteenth century, for example, American engineers developed their own theoretical works on the strength of materials, drawing on but modifying earlier scientific research. When engineers needed results that bore on their practical problems, they looked to engineering research, not pure science.

The idea that there are important traditions of technological knowledge has resonated with and influenced the thinking of other historians and philosophers of technology. Rachel Laudan's problem-solving model of technology (1984) assumes that the development of technologies is a research process, driven by interesting problems. The sources of problems Laucan identifies are: actual and potential functional failure of current technologies, extrapolation from past technological successes, imbalances between related technologies, and, much more rarely, external needs demanding a technical solution. It is notable that all but the last of these problem sources stem from within technological knowledge traditions. Edward Constant's (1964) Kuhnian model of normal and revolutionary technology adopts a very similar picture of technology as research, and a similar picture of engineering and other technological knowledge forming its own traditions.

As we will see in later chapters, for a group of people to have its own tradition of knowledge suggests that that knowledge will be tied to the group's social networks and material circumstances. There is a measure of practical incommensurability between knowledge traditions, seen in difficulties of translation. In addition, some knowledge within a tradition is tacit, not fully formalizable, and requires socialization for it to be passed from person to person.

So far, we have seen arguments for the autonomy of technological practice. A separate set of arguments challenges the picture of technology as applied science by insisting on the lack of distinctness of science and technology. For example, in a study of the idea of large technological systems, the historian of technology Thomas Hughes claims that "persons committed emotionally and intellectually to problem solving associated with system creation and development rarely take note of disciplinary boundaries, unless bureaucracy has taken command" (Hughes 1987: 64). "Scientists" invent, and "inventors" do scientific research—whatever is necessary to push forward their systemic program.

Actor-network theory's term *technoscience* suggests a pragmatic characterization of scientific knowledge. For the pragmatist, scientific knowledge is about what natural objects can be made to do. Thus Knorr Cetina's book *The Manufacture of Knowledge* (1981) argues that laboratory science is about what can be constructed, not about what exists independently. Peter Dear (1995) argues that the key to the scientific revolution was a new orientation to experimental inquiry. Even though it produced unnatural objects, experimentation became acceptable because it showed what could be constructed of the natural world. Patrick Carroll-Burke (2001) argues that this practical orientation has remained intact, that science is defined by its use of a number of different types of "epistemic engines," devices for making natural objects more knowable. For the purposes of this chapter, the pragmatic orientation is relevant in that it draws attention to the ways in which science depends upon technology, both materially and conceptually.

The term *technoscience* also draws attention to the increasing interdependence of science and technology. We might see it as odd that historians are insisting on the autonomy of technological traditions and cultures precisely when there is a new spate of science-based technologies and technologically-oriented science—biotechnologies, new materials science, and nanotechnology all cross obvious lines.

Latour's networks and Hughes's technological systems bundle many different resources together. Thomas Edison freely mixed economic calculations, the properties of materials, and sociological concerns in his designs (Hughes 1985). Technologists need scientific and technical knowledge, but they also need material, financial, social, and rhetorical resources. Even ideology can be an input, in the sense that it might shape decisions and the conditions of success and failure (e.g. Kaiserfeld 1996). For network builders nothing can be reduced to only one dimension. Technology requires heterogeneous engineering of a dramatic diversity of elements (Law; 1937; Bucciarelli 1994). A better picture of technology, then, is one that incorporates many different inputs, rather than being particularly dependent upon a single stream. It is possible that no one input is even essential, and could not be worked around, given enough hard work, ingenuity and other resources.

To sum up, scientific knowledge is one resource on which engineers and inventors can draw, and perhaps on which they are drawing increasingly. But there is no reason to see it as a dominant resource. Rather, the development of technology is a complex process that integrates many different things: different kinds of knowledge—including its own knowledge traditions—and different kinds of material resources. At the same time, it is clear that science draws on technology for its instruments, and perhaps also for some of its models of knowledge, just as some engineers may draw on science for their models of engineering knowledge. There are multiple relations of science and technology, rather than a single monolithic relation.

Box 1.1 Changing modes of research

Whatever the relations are between science and technology, they are shifting. For example, although corporate research has always been supposed to lead to profits, many companies created research units the goals of which were shielded from business concerns so that they could pursue open-ended questions. Since the 1980s, restructuring of many high-technology companies especially in the United States has led to the shrinking of those laboratories, and to their changing relationships with the rest of the company. One researcher comments that "I have jumped from theoretical physics to what customers would like the most" (Varmi 2000: 405). In part this is a change in the notion of the purposes of science, but it may also represent a change in people's sense of science's technological potential—as the idea of the "knowledge economy" suggests.

Going alongside this is a sense that universities, and academic research, are also changing. Universities, and university researchers, are increasingly patenting their results, and entering into partnerships with corporations to fund research and develop products. There have been a number of different formulations of the changing structures of research. Much discussed is the idea that there is a new "mode" of knowledge production (Gibbons et al. 1994; Ziman 1994). Instead of the discipline-bound problem-oriented research of mode 1, mode 2 involves transdisciplinary work, possibly involving actors from a variety of types of organizations, in which the application is clearly in view. Some of the same authors as put forward the mode 2 concept have developed an alternative that sees an increase in "contextualization" of science, a process in which non-scientists become involved in shaping the direction and content of specific pieces of research (Nowotny, Scott, and Gibbons 2001). There might be an increase in the importance of research that is simultaneously basic and applicable (Stokes 1997). Another alternative formulation of the change is in the more organizational terms of a "triple helix" of university-industry-government in interaction (Etzkowitz and Leydesdorff 1997). Governments are demanding of universities that they be relevant in trade for support, universities are becoming entrepreneurial, and industry is buying research from universities. The change might also be put in more negative terms, that there is a crisis in key categories that support the idea of pure scientific research: the justification for and bounds of academic freedom, the public domain, and disinterestedness have all become unclear, disrupting the ethos of pure science (McSherry 2001).

Critics of each of these formulations charge that the changes described are not nearly as abrupt as they are portrayed to be, but these criticisms do not take away from the feeling that dramatic changes are under way for scientific research, and that these changes are connected to the potential application of that research. Whether or not there is less pure science, there is certainly more applied science. Science and technology are seen as increasingly connected.

■ Does Technology Drive History?

A few of Karl Marx and Friedrich Engels's memorable comments on the influence of technology on economics and society can stand in for the position of the technological determinist, though they are certainly not everything that Marx and Engels had to say about the determinants of social structures. Looking at large-scale structures, Marx famously said: "The hand-mill gives you society with the feudal lord, the steam-mill, society with the industrial capitalist." Engels, talking about smaller-scale structures, claimed that "the automatic machinery of a big factory is much more despotic than the small capitalists who employ workers ever have been."

Technological determinism is the position that material forces, and especially the properties of available technologies, determine social events. The reasoning behind it is usually economistic: the available material resources, and the technologies for manipulating those resources, form the environment in which rational economic choices are made. In addition, technological determinism emphasizes "real-world constraints" and "technical logics" that shape technological trajectories (Vincenti 1995). Therefore, social variables ultimately depend upon material ones.

There are a number of different technological determinisms (see Bimber 1994), but the central idea is that technological changes force social adaptations, and through this narrowly constrain the trajectories of human history. Robert Heilbroner, supporting Marx, says that

> the hand-mill (if we may take this as referring to late medieval technology in general) required
> a work force composed of skilled or semiskilled craftsmen, who were free to practice their

occupations at home or in a small atelier, at times and seasons that varied considerably. By way of contrast, the steam-mill—that is, the technology of the nineteenth century—required a work force composed of semiskilled or unskilled operatives who could work only at the factory site and only at [a] strict time schedule. (Heilbroner 1994 [1967]: 60)

Because economic actors make rational choices, class structure is determined by the dominant technologies. This sort of reasoning applies to the largest scales, but also to much more local decisions. Thus technology shapes economic choices, and through those shapes history.

Some technologies appear particularly compatible with some types of political and social arrangements. In a well-known essay, Langdon Winner (1986a) asks: "do artifacts have politics?" He concludes that they do. Following Engels, he argues that some complex technological decisions will lend themselves to more hierarchical organization than others, in the name of efficiency—the complexity of modern industrial production does not lend itself to consensus decision-making. In addition, Winner argues, some technologies, such as nuclear power, are dangerous enough that they may bring their own demands for policing, and other forms of state power. And finally, individual artifacts may be constructed to achieve political goals. In a much-cited example, Winner describes how New York's "Master Builder" of roads, Robert Moses, designed overpasses on Long Island's parkways that were low enough to discourage buses, thus reserving Long Island's beaches for the car-owning classes. This example has been shown to be less straightforward than Winner originally portrayed it (Joerges 1999), but the point can be made with mundane examples. Speed bumps perform the political purpose of reducing and slowing traffic on a street, and thus increasing the property values of a family-oriented neighborhood; this technology becomes a form of long-distance control (Latour 1993a). To take a different type of example, David Noble argues that in the history of industrial automation choices were made to disempower key groups (Noble 1984). Numerical control automation, the dominant form, was developed to eliminate machinist skill altogether from the factory floor, and therefore to eliminate the power of key unions. While also intended to reduce factories' dependence on skilled labor, record-playback automation, a technology not developed nearly as much, would have required the maintenance of machinist skill, to reproduce it in machine form (for some related issues, see Wood 1982).

Even for non-determinists, the effects of technologies can form an important site of study. As we saw in chapter 1, a key part of the pre-S&TS constellation of ideas on science and technology was the study of the positive and negative effects of technologies, and the attempt to think systematically about these effects. That type of work continues, and is part of the more practical and applied portion of S&TS. At the same time, some researchers have been challenging a seemingly unchallengeable assumption, the assumption that technologies have any systematic effects at all! In fact, they challenge something slightly deeper, the idea that technologies have essential features. If technologies have no essential features, then they should not have systematic effects, and if they do not have any systematic effects then they cannot determine the structure of the social world.

No technology—and in fact no object—has only one potential use. Even something as apparently purposeful as a watch can be simultaneously constructed to tell time, to be attractive, to make profits, to refer to a well-known style of clock, to make a statement about its wearer, etc. Even the apparently simple goal of telling time might be seen as a multitude of different goals: within a day one might use a watch to keep on schedule, to find out how long a bicycle ride took, to regulate the cooking of a pastry, to notice when the sun set, and so on. Given this diversity, there is no essence to a watch. And if the watch has no essence, then we can only say that it has systematic effects within a particular human environment. Change that environment and one changes what the watch does.

Trevor Pinch and Wiebe Bijker (1987), in their work on "Social Construction of Technology" (SCOT), develop this point into a framework for thinking about the development of technologies. In their central example, the development of the safety bicycle, the basic design of most twentieth-century bicycles, there is an appearance of inevitability about the outcome. The standard modern bicycle is stable, safe, efficient, and fast, and therefore we might see its predecessors as important, but ultimately doomed, steps toward the safety bicycle. On Pinch and Bijker's analysis, though, the safety bicycle did not triumph because of an intrinsically superior design. Some users felt that other early bicycle variants represented superior designs, at least superior to the early versions of the safety bicycle with which they competed. For many young male riders, the safety bicycle sacrificed style for a claim to stability, even though new riders did not, of course, find it very stable.

Young male riders were one *relevant social group* that was not appeased by the new design. There is *interpretive flexibility* both in the understanding of technologies and in their design. We should see trajectories of technologies as the result of rhetorical operations, defining the users of artifacts, their uses, and the problems that particular designs solve.

On a SCOT analysis, the success of an artifact depends in large part upon the strength and size of the group that takes it up and promotes it. Its definition depends upon the associations that different actors make. Interpretive flexibility is thus a necessary feature of artifacts, because what an artifact does and how well it performs are the results of a competition of different groups' claims. Thus the good design of the safety bicycle cannot be the mover behind its success; good design is instead the result of its success.

Keith Grint and Steve Woolgar (1997) have developed this argument further. They take the metaphor of "technology as text," and try to show that technologies are interpretable as flexibly as are texts. They show, for example, that the Luddites of early nineteenth-century Britain adopted a variety of interpretations of the factory machines that they did and did not smash. Although some saw the factory machines as upsetting their preferred modes of work, others saw the problem in the masters of the factories. And, they argue, resistance to the new technologies diminished when new left-wing political theories articulated the machines as saviors of the working classes. The machines, then, did not have any single set of effects.

A study done by Woolgar shows some of the work needed to give technologies determinate meanings. Woolgar acted as participant-observer in a computer firm that was in the process of developing a computer, with bundled software, for educational markets. At a point well into the development process, the firm needed to test prototypes of their package, to see how easy it was for unskilled users to figure out how to perform some standard tasks. On the one hand these tests could be seen as revealing what needed to be done to the computers in order for them to be more user-friendly. On the other hand they could be seen as revealing what needed to be done to the users—how they needed to be defined, educated, controlled—to make them more computer-friendly: successful technologies require *configuring the user*. The computer, then, does what it does only in the context of an appropriate set of users.

But surely some features of technologies defy interpretation? We might, for example, ask with Rob Kling (1992): "What's so social about being shot?" Everything, say Grint and Woolgar. In a tour de force of anti-essentialist argumentation, Grint and Woolgar argue that a gun being shot is not nearly as simple a thing as it might seem. It is clear that the act of shooting a gun is intensely meaningful—some guns are, for example, more manly than others. But more than that, even injuries by gunshot can take on different meanings. When female Israeli soldiers were shot in 1948, "men who might have found the wounding of a male colleague comparatively tolerable were shocked by the injury of a woman, and the mission tended to get forgotten in a general scramble to ensure that she received medical aid" (R. Holmes, quoted in Grint and Woolgar 1997: 159). Even death is not so certain. Leaving aside common uncertainties about causes of death and the timing of death, there are cross-cultural differences about death and what happens following it. No matter how unmalleable a technology might look, there are always situations, some of them highly theoretical, in which the technology can take on unusual uses or interpretations.

To accept that technologies do not have essences is to pull the rug out from under technological determinism. If they do nothing outside of the social and material contexts in which they are developed and used, technologies cannot be the real drivers of history. Rather these contexts are in the drivers' seats. This recognition is potentially useful for political analyses, as particular technologies can be used to affect social relations (Hård 1993). This has been explored most in the context of labor relations (e.g. Noble 1984) and in the context of gender relations (e.g. Cockburn 1985; Cowan 1983; Kirkup and Smith Keller 1992).

There should, then, be no debate about technological determinism. However, in practice nobody holds a determinism that is strict enough to be completely overturned by these arguments. Even the strictest of determinists, like Heilbroner, admit that social forces play a variety of important roles in producing and shaping technology's effects. Such a "soft" determinism is an interpretive stance, according to Heilbroner, that directs us to look first to technological change to understand economic change. Choices are made in relation to material resources and opportunities. To the extent that we can see social choices as economic choices, technology will play a key role.

The anti-essentialist strain that has developed within S&TS is a counterbalancing interpretive stance. Anti-essentialists show us that even soft determinism must be understood within a social framework, in that the properties of technologies can be determinative of social events only once the social world has established what the properties of technologies are. It thus directs us to look to the social world to understand technological change and its effects. This is perhaps most valuable for its constant reminder that things could be different.

Just paying attention to anti-essentialist lessons would remove the interest in studying technology. We study technology because artifacts appear to do things, or at least are made to do things. Therefore we need a nondeterministic theory of technology that can accommodate the obduracy of artifacts. We need something like Wiebe Bijker's theory of *sociotechnology*, a theory of the thorough intertwining of the social and technical (Bijker 1995). This theory draws heavily on work on technology as heterogeneous engineering, and on Bijker's work with Trevor Pinch on SCOT. A key concept in Bijker's theory is that of the *technological frame*, the set of practices and the material and social infrastructure built up around an artifact or collection of similar artifacts (Bijker 1995: 123). As the frame is developed, it guides future actions. A technological frame, then, may reflect engineers' understandings of the key problems of the artifact, and the directions in which solutions should be sought. It may also reflect understandings of the potential users of the artifact, and users' understanding of its functions. If a strong technological frame has developed, it will cramp interpretive flexibility. The concept is therefore useful in helping to understand how technologies can appear deterministic, while only appearing so in particular contexts.

Even if technologies do not have essential forms—properties that they have independent of their interpretations, or functions that they perform independently of what they are made to do—essences can return, in muted form, as dispositions. For Francis Bacon, with whom we started this chapter, the reason that scientific research should translate into technological benefits was that scientific research investigated what substances could be made to do. The form of a substance, for Bacon, is its response to circumstances. The forms of substances are those underlying natures that express the potentialities of those substances. This is why Bacon says that "knowledge and human power are synonymous."

One might make the same argument about the properties of technological artifacts. Artifacts do nothing by themselves, though they can be said to have effects in particular social contexts. To the extent that we can specify the relevant features of their social contexts, we might say that technological artifacts have Baconian forms. Particular pieces of technology can be said to have definitive properties, though they change depending upon context. Material reductionism, then, only makes sense in a given social context, just as social reductionism only makes sense in a given material context.

Does technology drive history, then? History could be almost nothing without it. As Bijker puts it, "purely social relations are to be found only in the imaginations of sociologists or among baboons." But, equally, technology could be almost nothing without history. Bijker continues, "and purely technical relations are to be found only in the wilder reaches of science fiction" (Bijker 1995).

Box 1.2 Were electric automobiles doomed to fail?

David Kirsch's history of the electric vehicle illustrates both the difficulty and power of deterministic thinking (Kirsch 2000). The standard history of the internal combustion automobile portrays the electric vehicle as doomed to failure. Compared with the gasoline-powered vehicle, the electric vehicle suffered from lack of power and lack of range, and therefore could never be the all-purpose vehicle that consumers wanted. Kirsch argues, however, that "technological superiority was ultimately located in the hearts and minds of engineers, consumers, and drivers, not programmed inexorably into the chemical bonds of refined petroleum" (Kirsch 2000: 4).

Until about 1915 electric cars and trucks could compete with gasoline-powered cars and trucks in a number of market niches. In many ways electric cars and trucks were the more natural successors to horse-drawn carriages. Gasoline-powered trucks were faster than electric trucks, but for the owners of delivery companies speed was more likely to damage goods, to damage the vehicles themselves, and anyway was effectively limited in cities. Because they were easy to restart, electric trucks were better suited to making deliveries than early gasoline-powered trucks; this was especially true given the horse-paced rhythm of existing delivery service, which demanded interaction between driver and customer. Electric taxis were fashionable, comfortable, and quiet, and for a time were successful in a number of American cities, so much so that in 1900 the Electric Vehicle Company was the largest manufacturer of automobiles in the United States.

As innovators, electric taxi services were burdened with early equipment, they sometimes suffered poor management, and they were hit by expensive strikes. They also failed to participate in an integrated urban transit system that linked rail and road, to create a niche that they could dominate and in which they could then innovate. Meanwhile, Henry Ford's grand experiment in producing low-cost vehicles on assembly lines helped to spell the end of the electric vehicle. The First World War created a huge demand for gasoline-powered vehicles, which were better suited to war conditions than were electric ones. Increasing suburbanization of US cities meant that electric cars and trucks were restricted to a smaller and smaller segment of the market. Of course, that suburbanization was helped along by the successes of gasoline, and thus we might argue that the demands of consumers not only shaped, but were shaped by the technologies.

In 1900, then, the fate of the electric vehicle was not sealed. Does this failure of technological determinism mean that electric cars could be rehabilitated? According to Kirsch, that is unlikely. In 1900, both gasoline and electric cars fit the material and social contexts they faced, albeit very imperfectly. In the year 2000, material and social contexts have been shaped around the internal combustion engine. Given how unlikely it is that electric cars could compete directly with gasoline cars in these new contexts, their future does not look bright. Thus while a global technological determinism fails, within some contexts a soft determinism serves a useful heuristic purpose.

Historiographical Layers in the Relationship between Science and Technology

Reading 2

Abstract

This paper aims to examine historiographical layers in the historical narrative on the relationship between science and technology, a topic which has been exhaustively discussed without consensus by both historians of science and technology. I will first examine two extreme positions concerning this issue, and analyze the underlying historiographical standpoints behind them. I will then show that drawing implications for the relationship between science and technology from a few case-studies is frequently misleading. After showing this, I will move to the "macrohistory" of the relationship between science and technology, which reveals a long process in which the barriers between them gradually became porous. Here, I will examine the historical formation of three different kinds of "boundary objects" between science and technology, which facilitated their interactions by making their borders more permeable: instruments as a material boundary object; new institutions, laboratories, and departments as a spatial boundary or boundary space; and new mediators as a human hybrid.

■ Historiographical Layers in the Relationship between Science and Technology

Science/Technology Relationship: A Historiographical Introduction

No other issue in the history of science and technology has been so exhaustively debated without consensus than the relationship between science and technology. In these debates, two extreme positions have been clearly articulated. The first position is that science and technology have long been intimately connected and are almost fused into one, to such an extent that any distinction between them is now meaningless. It even argues that the separate terms—science and technology—should be replaced with one term, "technoscience." The second position, in contrast to the first, is that science and technology were, and largely still are, distinct activities, each with its own norms, methods, and communities. Between these two extreme opinions, there are a myriad of different opinions, constituting a wide spectrum on this issue.

Who takes up these two ends of the spectrum? Marxist historians and philosophers of science have favoured the idea that science and technology have been intimately connected. Benjamin Farrington, for one, argued that the beginnings of science in ancient Greece were closely associated with material production as can be seen among the pre-Socratic philosophers such as Thales, Anaximander, and Anaximenes, who were engaged in technical or manual activities. The relationship between them became remote as a result of the inception of the idealist natural philosophy of Plato and Aristotle. It was the Scientific Revolution in the sixteenth and seventeenth centuries that remarried science with technology. The philosophical foundation of the remarriage, Farrington argued, was best reflected in Francis Bacon's "industrial" philosophy of science. Other Marxists had similar opinions. In his very influential article, Borris Hessen claimed that Newton's new mechanics in *Principia* was caused by the technological and economic needs of the seventeenth-century European society, such as navigation and ballistics. J.D. Bernal's *Science in History* and *Science and Technology in the Nineteenth Century* also stress the intimate connections between science and technology throughout history. According to him, the contributions of science to technological developments, and the reverse contributions of technology to the progress of science became numerous since the Industrial Revolution. This made science an important part of the productive force of a society.

The Marxist idea was influential in the 1970s, but the recent notion of "technoscience" has little to do with the Marxist tradition. One of its intellectual origins is in Bruno Latour's and Michel Callon's actor-network theory, according to which scientific facts are inseparably linked with technological artifacts. The link between them has constructivist, and hardly Marxist, grounds. The actor-network theory suggested an extended definition of agencies to include non-human, as well as human, actors. According to this theory, scientific facts are a sort of artifact, in the sense that they are stabilized by the alliance between humans and non-humans (instruments and other non-human actors in the laboratory). The alliance is also an essential ingredient for the expansion of the technoscientific network outside the laboratory. The actor-network is thus formed among different groups of humans and non-human agencies. This renders a strict distinction between science and technology meaningless. That scientists are thinkers and engineers are tinkers, or that science and technology are two different human activities is an outmoded notion which should now be discarded.

Who takes the other extreme? A majority of historians of technology in North America has moved, either consciously or unconsciously, to this side. The establishment of the "Society for the History of Technology" (SHOT) in 1958 was initiated by a discord between some historians of technology and the then editor of *ISIS*, the official journal of the History of Science Society (HSS). During its early years, the history of technology, as a discipline, faced the task of defining its boundary between two relatively well-established fields, the history of science and economic history. For this, the SHOT historians of technology should undermine two prevalent arguments: 1) technological development is determined by science and 2) technological development is determined by economic need. Of the first, there was the "applied-science thesis" which argued that technology is merely a product of the application of scientific laws and methods. In connection with this thesis, the "linear model" of technological development proposed that basic science would always bring about technological invention and innovation.

Against this applied-science thesis and linear model, historians of technology presented two novel ideas. First, they maintained that scientists and engineers have different communities. Further, the communities of civil, mechanical, electrical, and chemical engineers have developed different, and inverted, values and norms than those of the scientific community. In the scientific community, for example, the abstract understanding of natural phenomena was given the highest credit, while abstract understanding was least valued in the engineers' community. Edwin Layton expressed that science and technology are the "mirror-image twins." They look like twins, but their appearances (their norms and values) are inverted, reflecting their "seek-to-know" versus "seek-to-do" imperatives. Secondly, historians of technology argued that the essence of technology lies in technological knowledge, rather than in the hardware. It has its own method, data, and content, but its essence consists in its pursuit of efficiency and its emphasis on design ability. On these two grounds, historians of technology proposed that the interaction between science and technology should not be understood as an application of scientific knowledge to technological hardware, but as an interaction between two equal sets of knowledge.

Historians of science never reached a consensus on this issue. However, historians like Alexandre Koyré, Rupert Hall, Richard Westfall, and Charles Gillispie, have pointed to the absence of direct, synergetic influence between science and technology until quite recently. Koyré characterized the Scientific Revolution by intellectual innovations such as "the destruction of the cosmos" and "the geometrization of space," for which the role of experiment and instruments, let alone technology, were minimized. Hall asserted that the impact of science upon technology during the Scientific Revolution was minimal. Technology and art infiltrated into science, not because the boundary between them became porous, but "scholars" in universities attended to and adopted the method of the "craftsmen" outside the university. Westfall criticized Farrington's concept of "industrial science," because, he says, what happened to the relationship between science and technology had little to do with economic sectors that can be properly categorized as "industrial." Hall argued that the direct interaction between technology and scientific knowledge was almost absent even during the Industrial Revolution. Important technological developments of the day, such as the textile machines, the steam engine, railway, the use of charcoal for the production of iron, and the chemical production of soda, were not made possible by science. Taking examples from the French *Encyclopaedia* movement, Gillispie asserted that science provided technology with a rational classification and systematization of existing techniques and industrial procedures—a process which he called the "natural history of industry."

There are, however, interesting differences between historians of science like Hall and Gillispie, on the one hand, and the SHOT historians, on the other. Hall and Gillispie took for granted that *modern* technology is deeply indebted to science. Hall apparently thought that modern technology was a sort of applied-science when he listed the four contributions of science to technological knowing: 1) mathematical analysis; 2) controlled experiments; 3) natural laws such as thermodynamics and genetics; and 4) new natural phenomena. Thus, modern technological knowledge is "closely approximate to" modern science in basic ideas. This is what most, perhaps not all, of the SHOT historians could not have accepted. According to the SHOT historians, it was not the passive adoption by technology of high scientific theories, but an interactive process between them that led to the birth of a new engineering science. Engineers have formulated systematic theories even when no scientific guidance was available. On the basis of detailed studies into aeronautical engineering, Walter Vincenti has unmasked the differences in, for example, the uses of models and experiment in science and engineering. While scientists use scale models and do experiment to build or test hypotheses, engineers use models and experiment to obtain data for design, with which to bypass the difficulty caused by the absence of an exact quantitative theory.

The issue of the relationship between science and technology has become more complicated due to conflicting interpretations of historical case studies. The Marxist historians, the SHOT historians, and the historians of science used historical examples and case studies to make their points more convincing. These case studies have not apparently settled the long controversy over the relationship between science and technology. But why? Why such examples and case-studies cannot settle the controversy? Why do historians make different interpretations out of the same historical event? The next section will examine these issues. I will show that the microhistorical examinations of case studies will lead to different, and even conflicting, conclusions on science/technology relationship.

Historical Case-Studies and the Limit of Implications

Those who supported the idea of the distinctiveness between science and technology preferred to refer to Project Hindsight—a project conducted in the early 1960s with support from the Department of Defence of the United States. Project Hindsight probed the contribution of science and technology to the development of the modern military defence system. Investigators had first sorted out "events" that led to the system, and apparently discovered that only "0.3% of these events" were caused by science. More than 90% of these events turned out to be of technological origin, and 8% of applied-science origin. Yet, people who disagreed with the findings of the Project Hindsight preferred Project TRACES, a counter-project sponsored by the National Science Foundation. Project TRACES traced the origins of several innovations like contraceptives, the electron microscope and video tape recorders, and showed that they were rooted in basic (non-mission oriented) scientific research. As we can easily see, these two different conclusions had different policy implications.

Each group tends to emphasize different aspects of the same historical event. Thomas Edison's inventions, and in particular his electric lighting system, have been stressed as showing the autonomy of technology from science. However, discontents have stressed the crucial contribution of the physicist F. Upton, Edison's employee, to the development of Edison's electric lighting system. The engineering method of W.J.M. Rankine (1820–1872), one of the pioneers of modern engineering, has been interpreted either as evidence showing the application of science to engineering or as evidence showing the formation of a new field of engineering distinct from science. The invention of the Diesel engine by the German engineer Rudolf Diesel has been depicted either as showing the contribution of science (thermodynamic principles such as Carnot's cycle) to technology or the uselessness of scientific ideals (again Carnot's cycle) in the real world of technology. The invention of the laser was regarded as showing either a fruitful interaction between science (quantum physics) and microwave engineering or a huge gap between scientific ideas on stimulated emission (suggested in the 1910s) and its application to the laser (made in the late 1950s). Faraday's discovery of electromagnetic induction in 1831 has been interpreted as a crucial piece of evidence showing the contribution of science to technology, since it made possible the electric motor and dynamo. Yet, those who are inclined to oppose this implication would say that it was not Faraday, but many obscure technologists, who essentially contributed to the invention of the commercial dynamo which was only available at the end of 1860s, thirty years after

Faraday's discovery. Likewise, the entire history of electricity in the nineteenth century has been described in two different ways. A. Keller asserted that "without Volta, Davy, Oersted, Ohm, Ampere and Faraday, there would have been no telegraph, telephone, electrolysis, electroplating, no batteries, generators, illumination or trams." Contrarily, Derek de Solla Price characterizes the same period as the period when a "revolutionary change in the technology of science caused repercussions all over science and technology and created entire sets of new methodologies."

Why do these historical interpretations diverge? There are several reasons for this. First, a disciplinary gap between the history of science and technology certainly contributed to the worsening of the situation. Historians of science seldom examine "inside technology," while historians of technology sometimes hold an outmoded conception of science. As de Solla Price once complained, "our parascientific professions happen to have been professionalized quite separately so that most historians of science get a certificate of ignorance in the history of technology and vice versa." Secondly, there is a semantic problem. As Otto Mayr pointed out, the term science or technology is a vague "umbrella term", which indicate "general areas of meaning without precisely defining their limits." Other terms such as (pure or practical) engineering, engineering science, scientific engineering, technological knowledge, and theory and practice, as well as various concepts used to denote their relationships such as interaction, application, flow of information, mediation, and transformation, cannot be precisely defined. Metaphors and models such as "dancing pair," "mirror-image twin," "symbiosis," and "dialectic interactions" distorts as much as reflects the relationship between science and technology. Third, historians' general propositions as to the relationship between science and technology, consciously or not, reflect both their analysis of specific past case-studies and their present concerns. I think that Layton's and other SHOT historians' ideals on the distinctiveness between scientific and technological knowledge reflects the ideas of some nineteenth-century and twentieth-century American engineers who stressed for the professionalization of engineering the independence of technology from science. Let me give another example. In the history of radio, Marconi developed his own version of the history of wireless telegraphy in which technology was largely independent of science, while the British Maxwellians such as Oliver Lodge undermined it by stating various contributions of science to early radio. The actors' histories were inscribed in their scientific articles and monographs, memoirs, autobiographies, obituaries, and patents, and in archives such as letters and unpublished diaries. The interpretations of later historians were heavily influenced by these sources, which they regarded as "authority."

So, among these competing conclusions, which one does reflect the true historical relationship? Can we draw an "objective" picture that truly represents the relationship between science and technology? What I have tried to show in this section is that drawing a general conclusion on the relationship between science and technology out of a single or a series of specific historical case-studies is often risky, for one can draw multiple, and even conflicting, implications from the same historical event. In the next section, I will make one attempt to bypass this difficulty. For this, I will propose a "macrohistory" of the relationship between science and technology, in which long-term trends and patterns are analysed. The macrohistory that I will present will emphasize the historical formation of three different kinds of boundary objects between science and technology, which made their boundary more porous, and which facilitated their interactions. They include: instruments as a material boundary object; new institutions and laboratories as a spatial boundary object or boundary space; and new group of people who mediated between science and technology as a boundary person or a human hybrid.

Boundary Objects and Macrohistorical Relationships between Science and Technology

Science (or natural philosophy) and technology (or arts) were rather distinct and separate fields in ancient times and in the Middle Ages. There were several reasons for this separation. First, natural philosophy and technology were carried out by different groups of people. Natural philosophy was done by scholarly intellectuals, while technology was carried out by uneducated artisans who usually belonged to the lower class. Because of this, we do not know the names of the inventors of the grinding stone, siphons, water pumps, the mechanical clock, and the wind-mill, all designed in ancient and medieval times. Secondly, natural philosophy and technological activity were considered to be quite different in nature. The goal of natural philosophers was to discuss natural phenomena for philosophical or theological purposes, while the goal of artisans invent useful things. Aristotelian natural philosophy, which prevailed in the medieval ages, took for granted a strict dichotomy between the natural and the artifactual, which justified natural philosophers' low appreciation of technological activity.

The *macrohistory* of the relationship between science and technology reveals a long process in which the barriers that had blocked the interaction between science and technology gradually became porous and partially removed. The trigger for this was prepared in the Renaissance when the social status of engineers rose considerably. The so-called Renaissance engineers made fortune and established social prestige by constructing buildings, ships, and canals. Further, they educated themselves to be more erudite. They contacted university scholars, learned classical languages like Greek and Latin, read Euclid and Archimedes, and wrote books on engineering topics. This brought university scholars to become interested in engineering problems. Simon Stevin and Galileo tackled statical and mechanical problems such as those in the lever, buoyancy, and the projectile. Pascal and Torriceli tried to explain hydraulics and phenomena in water pumps. Others examined geometrical problems in military fortification, navigation and cartography. This naturally created several contact points between scholars and engineers.

Material Boundary Object: Instruments

Another change during the Scientific Revolution was the introduction of an experimental method into science. Aristotle considered observation to be important to science, but was strictly against the employment of experimental methods, because he thought that experimental intervention would disturb the true course of nature. Once experiment was introduced into science by Francis Bacon, Galileo, and Robert Boyle, and accepted as the proper scientific method, technologies rushed into science in the form of instruments. Scientists used and constructed various instruments such as inclined planes, pumps, pulleys, levers, lenses, prisms, and clocks for their scientific research. Telescopes and microscopes greatly widened human senses, generating a series of epistemological controversies over the authenticity of the images they produced. Robert Boyle's air pump created a vacuous space which did not normally exist in nature. The natural world that natural philosophers were interested in was no longer "naked" nature, but a sort of artifactual world—or "nature in the wider sense"—created or mediated by instruments. By doing so, instruments shortened the epistemological gap between science and technology.

I would argue that instruments such as the microscope, telescope and air-pump, technologies such as the pendulum clock, and demonstration machines such as Atwood's machine (eighteenth-century) provided the first interface for natural philosophers, instrument makers, and technologists. In other words, they became the first boundary object which lay overlapped across science and technology. Instruments mediated the interaction between science and technology in four ways. First, they constituted a major inflow of technology into science. Second, new artifacts created a new, strange effect that drew the attention of scientists to provide a scientific explanation. Scientists sometimes transformed such an effect into a standardized scientific technique or a marketable instrument. Lamps exhibited a strange effect named the Edison effect, and the Edison effect was transformed into the thermionic valve. Another example is a strange electrical property of some crystals. This was used as the crystal rectifier (the cat's whisker) in early wireless telegraphy. It then attracted the endeavour of several theoretical and experimental physicists in the first half of the twentieth century, which eventually led to the birth of the contact semiconductor in 1947. Third, a scientific theory designed for instruments could later be applied (with modifications) to machinery, as Maxwell's electromagnetic theories on electrical instruments were later used for heavy AC machineries. Finally, as Layton has asserted, measuring instruments (like the Prony dynamometer) designed to measure the efficiency of a technology (like the water turbine) can incorporate relevant scientific principles. This kind of instrument plays a crucial role for the diffusion and improvement of the technology.

Spatial Boundary Object or Boundary Space

Let's move to the second boundary object. The second boundary object was the social space in which scientists and engineers could meet and discuss their common interests. Since the late seventeenth century, coffeehouses, salons, and pubs in London were developed as a place where Newtonian natural philosophers like Desagulier and businessmen like James Brydges met and discuss on Newtonianism, navigation, commerce, insurance, and the newly invented Savory engine. In the last quarter of the eighteenth century, "Philosophical and Literary" Societies were established in various industrial cities such as Birmingham, Manchester, Leeds,

Liverpool, Newcastle, Sheffield, and Derby. The members of these societies included notable scientists, engineers, businessmen, and industrialists of the cities. The Lunar Society of Birmingham, the most famous of these societies, had as its members James Watt, Josiah Wedgewood, Joseph Priestley, Erasmus Darwin, and local industrialists and engineers. They discussed diverse scientific and technical issues such as the phlogiston theory of combustion, thermodynamics, and the steam engine. They believed that scientific knowledge would bring about technological and industrial advances.

Debates have taken place over whether scientific knowledge discussed in these societies really contributed to technological development. Neil McKendrick argued that Josiah Wedgewood, a successful ceramic industrialist and a member of the Lunar Society, was in fact much indebted to the knowledge and techniques of chemistry for his success in making china. But his critics would argue that Wedgewood's chemical knowledge was based on Joseph Priestley's theory of the phlogiston that proved to be non-existent, and further that Wedgewood simply legitimized his success in terms of the current scientific theory. This issue becomes more complicated, since historians have implicitly projected their present standard of science upon the past. Historians tend to think that an incorrect theory such as Priestley's phlogiston theory cannot result in successful applications. Wedgewood's knowledge, if successful, must have come from somewhere other than faulty science—for instance, from his accumulated experiences in china-making in the field. This kind of logic leads to the conclusion that sciences can be genuinely applied to technology only when their theories are correct. When did scientific theories obtain a self-correcting mechanism? One is likely to say that physical sciences in the second half of the nineteenth century, armed with mathematics and experiment, were quite different from those in the second half of the eighteenth century. Take Kelvin's thermodynamics, Maxwell's electromagnetic theory, or Liebig's organic chemistry and compare them with the eighteenth-century caloric theory, Galvanism, and the phlogiston theory. No wonder the new thermodynamics, electromagnetism, and organic chemistry—all of which constitute what we now call "science"—found their ways into the industrial and technological applications of science.

I have discussed the creation of new social spaces such as the literary and philosophical societies in which scientists, engineers, and industrialists interacted and talked about their common interests. These interactions became an intellectual motive force for the establishment of several provincial colleges such as Owens College in Birmingham and Mason College in Manchester which specialized in training local elites including industrialists and engineers. The social gap between university-based scientists, and self-employed inventors and consultant engineers, became almost negligible in these newly created colleges. In 1840, the first chair for an engineering professorship was established in the University of Glasgow, which accelerated the already existing interactions between scientists and engineers. In 1832, the British Association for the Advancement of Science (BAAS) was established, and in the 1840s it set up a committee for the purpose of exploring the magnetic field of the world. In 1860, telegraphers and physicists who had contacted in both the BAAS and submarine telegraphy companies agreed on establishing the Committee for Electrical Standards in the BAAS. Under this committee, theorists like James Clerk Maxwell worked with the practical telegrapher Fleeming Jenkin to find a way to determine the new unit and standard system. Through this collaboration, Maxwell obtained a new insight into the ratio of electrostatic to electromagnetic charge of electricity, which later became a basis of his electromagnetic theory of light.

Another important boundary space, the research laboratory in industry, was established as the belief in the mutual benefit between science and technology became popular and the actual contact between academics and industrialists became increasingly commonplace. In the field of dyestuff chemical industry, the university-trained chemists had established successful business of their own, as can be seen in the successful career of William H. Perkin who, as a student of A. Wilhelm Hofmann, synthesized the first chemical dye, aniline purple, in 1856 and became a successful businessman. In the 1870s, chemical industry already hired university-trained chemists for the problems of cost and quality. In the 1880s, the Bayer Company hired a few Ph.D. chemists to conduct research only. This became the germ for the Bayer Company's well-known research division, which soon found new colours for synthetic dyes. This led to the expansion of the division into a research laboratory. This chemical laboratory became a model for the electrical industry. In 1900, the General Electric Company in the United States established the first research laboratory among electrical industries under the directorship of W.R. Whitney, and hired physicists and chemists for long-term research. These scientists essentially employed a hybrid method characterized by the "natural study of the artificial" and "technological theories." Langmuir in the lab was even awarded the Nobel Prize in chemistry for his research that he performed in the laboratory. The Bell Telephone laboratory, which was established for solving serious technical problems in long-distance telephony, became another very successful company laboratory. Its contributions to pure science include C.J. Davisson and L.H. Germer's discovery of electron diffraction in 1927, and A. Penzias and R. Wilson's discovery of cosmic background radiation in 1962, each of which essentially contributed to physics and astronomy, respectively.

Boundary People and Engineering

The proliferation of the boundary spaces went hand in hand with the formation of hybrid people who were interested in, and contributed to, both science and technology. James Joule, one of the discoverers of the principle of energy conservation, was an early protocol. His experience in engineering research on electric battery and the steam engine made him adopt the language of engineers such as "duty" or "work," and this linguistic resource became an important ground for his discovery of the mechanical equivalent of heat. From the 1840s on, hybrid figures appeared frequently. William Thomson (later Lord Kelvin) worked on both electromagnetic theory and submarine telegraphy, combining theory and practice into one. His mathematical theory on the transmission of telegraphic signals became a model for scientific electrical engineering. In the field of civil engineering, W.J.M. Rankine integrated both theory and practice. In telephone engineering of the 1880s, Oliver Heaviside, who discovered the mathematical condition for the distortionless transmission of signals, followed the route that Thomson had paved thirty years previously to combine new mathematical techniques with the real world of technology. In AC power engineering, John Hopkinson, John Ambrose Fleming, and William Ayrton built the new discipline. These pioneers mediated between scientific theory and practical knowledge in the field, constructing the core of engineering knowledge.

The historical evolution of various engineering disciplines in the nineteenth and twentieth centuries is beyond the scope of this paper. However, I will briefly discuss three features of engineering which have been paid less attention by historians of both science and technology. First, historians have pointed out that engineering (or scientific engineering or engineering science) has mediated between science and technology, but they have implicitly assumed that science represents a high theory and technology represents practical know-how, and that the mediation has therefore taken place as the transfer of information from high to low. This assumption is sometimes not compatible with the historical fact that many scientists were neither interested nor contributed much to technology or that, at many times, their contributions were entirely neglected by practical engineers. In my articles on the history of electrical engineering, I proposed that mediation between science and technology in the early phase of electrical engineering was made possible by two kinds of "mediating people"—the scientist-engineers who, on the basis of their scientific training, understood the operation of machinery in scientific terms, and the practicing-engineers who systematized their accumulated knowledge in the field and in the workshop. I also stressed that the mediation took place on four different levels: that is, between a scientific and a "workshop theory"; between a scientific theory and a practice in the field and in the workshop; between a scientific practice in the lab and a workshop theory; and between a scientific practice in the lab and a practice in the field and the workshop.

Secondly, if we consider the creation and the development of, for example, electrical power engineering as a product of intense interaction between science and technology (each of which had its own theories and practices) through the mediation by hybrid engineers, we can divide its history into two qualitatively different periods. During 1880–1900, which I want to call the "formative period," the boundary of power engineering was highly flexible with no clearly defined principles and practices. The boundary was continuously redefined by the extensive interaction between the scientist-engineers and the practicing engineers. In this process, they actively drew their theories and practices from many resources including scientific theory, tacit knowledge acquired from craft skill, operating knowledge of machinery, and workshop as well as laboratory practices, and moulded them into electrical engineering. Science penetrated the real world of technology through this truly multidimensional interaction. However, during the period from 1900 on, which I want to call the "stabilized period," during which major problems were largely solved and the boundary of power engineering became rather rigid, engineers were more concerned with the increase in the efficiency of the system and the perfection of it. At this stage, they drew mainly on existing, well-established body of engineering theory and practice. The interaction between science and technology becomes a routine interaction between two well-defined sets of knowledge, that is, between science and engineering knowledge.

The third feature is more speculative. Historians have pondered over why the so-called science-based industry, which emerged in the second half of the nineteenth century, started with chemical and electrical science and engineering, which were then newly established, rather than with older, well-established engineering fields such as mechanical, civil, architectural, or hydraulic engineering, or with traditional sciences such as mechanics and astronomy. Think of the early phase of electrical engineering. In this case, the major resource that engineers could draw on was science. In the late 1870s and early 1880s, for instance, various scientific theories on direct and even alternating current phenomena were available. Various theories and hardware of measuring instruments, which had been devised for laboratory purposes, were at the hands of engineers. Science was the most readily available resource for engineering problems, since there was no established engineering theory and practice. It was not the epistemological superiority of science, but the efforts of hybrid engineers to mobilize science as a resource to solve particular problems, that started the science-based industry or engineering.

Concluding Remarks

The emergence and the spread of three boundary objects—instrumental, spatial, and human—that gradually blurred the distinction between science and technology was not smooth or straightforward. When we turn to *microhistories*, the apparently regular patterns and trends in the macrohistory dissolve into fuzzier, chaotic configurations. These microhistorical configurations seems to have been conditioned by various contingent and local factors which were seldom taken into account in the macrohistory. Tensions and conflicts between science and technology can always be found in microhistorical analyses. Even in such a hybrid space as the BAAS's Electrical Standards Committee, there was tension between engineers and scientists. Scientific engineering appears less clear when we examine it through a "microscope". William Thomson's theoretical work on signal transmission were in fact less essential to the success of the submarine telegraphy than innovations of a technological kind such as the improved insulation of the cables. Heaviside's innovative idea on distortion-less signal transmission was almost completely ignored by engineers at the time. The status of research in the corporate laboratory has been unstable due to the tension between managers who wanted to direct scientists to particular problems in the industrial production and scientists who wanted to maintain their autonomy over research.

These microhistories, however, cannot overrule the macrohistorical patterns and trends, any more than the macrohistory should not make us blind to local and temporal irregularities and contingencies. What these microhistorical irregularities and contingencies reflect is the essential instability inherent in boundary objects. The flexibility of a boundary object induces various, and even conflicting, elements as its constituents, but the same flexibility contributes to its instability. However, this instability can be a resources for its dynamics. A flexible, unstable boundary object is like a motion of the bicycle: the instability remains constructive as long as it is pushed forward. As long as a boundary object begets other boundary objects, it can maintain a dynamic instability. Some economists described the history of the capitalism in terms of dynamic instability in the sense that the inherent instability in capitalist economy remains constructive as long as it remains dynamic. This dynamic instability is a homology between the science-technology hybrid and the capitalist context in which this hybrid evolved.

Why in America?

Reading 3

Nathan Rosenberg

Let me try to clarify just what question it is I am trying to answer. I am not going to discuss the high overall rate of economic growth in the United States after 1800, that is, the rate of increase of per capita income. In fact, the rate of growth was not particularly high by comparison with later experience, although many scholars believe that it accelerated during the 1830s. If we extend our time horizon well back into the colonial period, we find that the rate of growth of per capita income between 1710 and 1840 did not exceed one-half of 1 percent per year.[1] Nor are we considering the overall rate of technological innovativeness which characterized the American economy, although that rate was doubtless very high. In spite of continued obeisance to the idea of Yankee ingenuity, it cannot be overstressed that America in the first half of the nineteenth century was still primarily a borrower of European technology. Although the rate of technical change was indeed high, most of the new technologies were not American inventions. Americans were rapid adopters of foreign technologies when it suited their economic needs, and they were also skillful in modifying someone else's technology to make it more suitable to their needs. They made abundant albeit selective use of European innovations in power generation, transportation, and metallurgy.[2] It is worth noting, moreover, that high rates of technological improvement were not always a part of the American experience. Such improvements seem to have played an insignificant role in the American economy in the eighteenth century. The latest scholarship on the colonial period suggests that the rate of inventive activity was very low, and that the slow but steady growth in productivity was dependent upon a cumulatively powerful combination of forces centering upon improvements in the organization of industries and the more effective functioning of markets and ancillary institutions.[3]

The question I propose to address, then, is not the high rate of technological change in America when that rate indeed became high in the nineteenth century. Rather, the question is one of direction and character. By the time of the Crystal Palace Exhibition in London in 1851, there was something so distinctive about many American goods that the British coined the phrase "the American system of manufactures." The expression was used to describe goods which were (1) produced by specialized machines, (2) highly standardized, and (3) made up of interchangeable component parts.[4] I propose to address myself to the question of causality.

Originally published in Otto Mayr and Robert C. Post (eds.), *Yankee Enterprise, the Rise, of the American System of Manufactures*, Smithsonian Institution Press, Washington (DC), 1981. The focus here on demand- and supply- side factors to explain the particular direction of American technological development complements the discussion of technological change in the energy and forest products sectors in chapters 9 and 12, respectively.

[1] Robert Gallman, "The Record of American Economic Growth," in Lance Davis, Richard A. Easterlin, and William N. Parker (eds.), *American Economic Growth,* Harper & Row, New York, 1972, p. 22.

[2] Nathan Rosenberg, *Technology and American Economic Growth,* Harper & Row, New York, 1972, chapter 3.

[3] See James Shepherd and Gary Walton, *Shipping, Maritime Trade and the Economic Development of Colonial North America,* Cambridge University Press, Cambridge, 1972.

[4] The degree or ease of interchangeability varied between products and over time. The degree of precision required for interchangeability varied considerably from one product to another. As one nineteenth-century commentator observed: "The more prevalent modern idea of the interchangeable in mechanism supposes a super-refinement of accuracy of outline and general proportions that is not always necessary or even desirable. It would be a criminal waste of time and substance to fit a harrow tooth with mathematical accuracy, but yet any harrow tooth should have a practical interchangeable relation to all harrows for which it is designed. The instructive rewards

Why did this system emerge first in America? By midcentury it was apparent that there was an important class of goods which was being produced in a distinctive way in America.[5] It is important, however, that the distinctiveness should not be overstated, nor should it be thought that America had also assumed a role of broad technological leadership across the whole manufacturing sector. That was far from the case. Indeed, even in gunmaking, commonly regarded as the main triumph of the American system of manufactures, American barrelmaking may still have lagged well behind Britain at midcentury.[6]

Now, if you call upon an economist to try to explain a particular phenomenon, such as why the technology which we have come to call the American system of manufactures first arose in the United States and not elsewhere, you are not entitled to express surprise over an explanation couched in terms of supply-and-demand analysis. That is what I propose to give, and I do so without apology. Presenting an historical explanation in terms of supply-and-demand forces does not *necessarily* mean that I am committed to some form of economic determinism. Supply and demand are nothing more than convenient conceptual categories which enable us to think about an event or process in a more systematic way and to organize our analysis in a way which lends itself more readily to an understanding of cause–effect relationships. The forces *underlying* these economic categories may be social, geographic, technological, or ideological, but such forces have economic content and effects, and I am particularly interested in focusing upon the economic implications or consequences of these forces. There is a difference, which I trust that I need not belabor, between *translating* certain phenomena into their economic consequences and insisting that economic variables are all powerful.

I have already suggested a distinction between the rate and direction of technological progress. However, these two factors are not as easily separable as may be thought at first glance. The point is that differences in the resource endowment and demand conditions of an economy go a long way toward determining what kinds of inventions—with what kinds of product characteristics and factor-saving biases—it will be profitable to develop and exploit. To the extent that technological progress is responsive to economic forces—and I would agree that it is highly responsive and needs to be understood in these terms—the inventions actually brought forth in a particular country will tend to be compatible with those special needs. We need to distinguish here between invention and adoption. It is obvious that only those inventions which are compatible with a country's needs will be widely adopted. I am making here the stronger assertion that a high proportion of the *inventions made* will also reflect the peculiar needs of the economic environment in which they are developed.

If, at any given time and state of technical knowledge, certain *kinds* of inventions are easier to create, for whatever reason, economies where such inventions are also economically appropriate would be expected (*ceteris paribus*, as economists are fond of saying) to find it easier to generate new and improved technologies. If economic forces tend to push economies with different factor endowments in different directions in the attainment of technological progress, and if the technical problems which have to be overcome are more difficult in some directions than others, then the rate of technological progress may not be independent of the direction in which the economy is being pushed.

I have developed this argument because I believe it has merit. The American experience in the nineteenth century involved an economy with resource endowment and demand conditions that pushed it in a direction somewhat different from the one that prevailed at the time for the economies of western Europe. This turned out to be a direction in which the inventive payoff was rather high. You may, if you wish, take this as a confirmation of the belief that a benevolent Providence watched over the affairs of America in the nineteenth century. Or you may simply regard it as luck or as the consequence of a particular resource endowment.

Whatever the label, I want to argue that what we call the American system of manufactures was a part of a larger process of economic adaptation, or even, if you prefer, economic evolution. American economic and

of folly would certainly overtake him who should attempt to make ploughshares and coulters with radical exactness; nevertheless, these essential parts of ploughs should be interchangeable among all ploughs to which they are adapted." W. F. Durfee, "The History of Interchangeable Construction," *American Society of Mechanical Engineers Transactions*, 14 (1893), p. 1228.

[5]The role of the manufacturing sector had increased very sharply during the decade of the 1840s. According to Gallman's estimates the contribution of manufacturing to national commodity output grew from 17 percent to 30 percent during the decade 1839–1849. Robert Gallman, "Commodity Output, 1839–1899," in the Conference on Income and Wealth's *Trends in the American Economy in the 19th Century*, Princeton University Press, 1960, p. 26.

[6]Paul Uselding, "Henry Burden and the Question of Anglo-American Technological Transfer in the Nineteenth Century," *Journal of Economic History*, 30 (1970), pp. 312–337.

social conditions in the nineteenth century pushed the search for new technologies in a specific direction. It was a direction which, both for social and economic reasons, was less appropriate in Europe. And, as it turned out, it was a direction in which there happened to exist a rich layer of inventive possibilities. In this sense, I will argue, the American system of manufactures was a species of a larger genus and, in this sense also, the high *rate* of technical progress was not entirely independent of the *direction* which that progress took.

The notion that the American system of manufactures requires explanation is reinforced by the facts that (1) the United States was, in the first half of the nineteenth century, still largely a technical borrower rather than a pioneer, and (2) technical innovations incorporating some of the elements of the system had been introduced earlier in Britain, mainly in the Portsmouth naval dockyards, but were not widely adopted elsewhere in that country.

■ Demand

What were the factors on the demand side that were so conducive in early-nineteenth-century America to the emergence of the American system of manufactures? There was a very rapid rate of growth in the aggregate demand for certain classes of commodities, which encouraged entrepreneurs to devise or adopt new production processes whose profitability required very long production runs. Adam Smith long ago taught that "the division of labor is limited by the extent of the market." The proposition applies to the use of specialized machinery as well as to specialized men. The fixed costs embodied in special-purpose machinery are warranted only if they can be distributed over a large volume of output. Therefore the confident expectation of rapidly growing markets is an extremely powerful inducement. Furthermore, the use of highly specialized machinery, as opposed to machinery which has a greater general-purpose capability, is contingent upon expectations concerning the *composition* of demand—specifically, that there will be strict and well-defined limits to permissible variations. Both of these conditions—rapid growth in demand and circumstances conducive to a high degree of product standardization—were amply fulfilled in early-nineteenth-century America.

What forces were responsible for these conditions? Probably the most pervasive force of all was the extremely rapid rate of population growth, primarily from natural increase but with immigration assuming a role of some significance in the 1840s. Between 1790 and 1860, the American population grew at a rate of nearly 3 percent per year (immigration included), a rate more than twice as high as that achieved by any European country even for much shorter periods. American fertility levels at the beginning of the nineteenth century were close to fifty per thousand—again, a rate far higher than European levels—and crude death rates were somewhere around twenty per thousand. The factors underlying this rapid expansion in population are many, but most important was probably the abundant supply of high-quality land in a society which was still predominantly agricultural. The abundance of good land, and the concomitant optimism about the future, were conducive to high fertility levels as well as an inducement to immigration.[7]

Although scholars differ concerning the long-term trend in the rate of growth of per capita income during the nineteenth century, there can be no doubt that the strong rate of population growth resulted in an exceedingly rapid rate of market growth. As early as 1840 the net national product of the United States was two-thirds or more of the size of Great Britain's.[8] In subsequent years America's net national product grew far more rapidly than Britain's.

Rapid population growth resulted in a very high rate of new household formation and therefore a rapid rate of growth in the demand for a wide range of manufactured commodities. This growth in market size was reinforced by other developments. Improvements in transportation, especially after the beginning of the canal-building period in 1815 and the beginning of railroad construction in the 1830s, served to link industrial centers with remote potential markets that previously had been largely self-sufficient. Thus, in many cases industrial producers and consumers were in a position to establish economically feasible market relationships with one another for the first time.

Many of the distinctive characteristics of the American market can be properly appreciated only in the context of its predominantly rural and agricultural character. Over 80 percent of the American labor force was in

---

[7]For an excellent, concise treatment of long-term demographic trends in America, see Richard A. Easterlin, "The American Population," chapter 5 in Davis, Easterlin, and Parker (eds.), *American Economic Growth*.

[8]Gallman, "Commodity Output, 1839–1899," p. 33.

agriculture in 1810 and this remained true for well over 60 percent in 1840.[9] The predominance of the agricultural sector in the nineteenth century, and the abundant endowment of natural resources more generally, served to shape the American environment in numerous ways which made it more receptive to the American system of manufactures. First of all, we have already noted how the rich and abundant supply of agricultural land contributed to high fertility levels and therefore to a rapid growth in the market. In addition, the abundance of land conferred another advantage upon the American economy, for it meant that food prices were relatively low. As a result, for any given income level or family size, there was a larger margin left over for the purchase of nonfood products, including those produced under the American system of manufactures. Thus the structure of relative prices in the United States, as a result of its resource endowment, was distinctly more favorable than in Europe for the emergence of manufactured goods. Both American farmers and urban industrial workers spent less of their income on food than did their European counterparts. Empirical evidence in support of this hypothesis has been carefully marshalled and analyzed by Albert Fishlow, relying upon data collected for the period 1888–91.[10] Since American food prices were, relatively, lower earlier in the nineteenth century, the case for the more favorable American conditions is even stronger.[11]

Yet another advantage conferred by America's natural-resource abundance was the easy accessibility to land ownership and the relatively egalitarian social structure which consequently emerged—with the major exception, of course, of the southern plantation system. Outside of the south there had emerged in colonial times and later a rural society which was far less hierarchical, and with property ownership far less concentrated, than in Europe. American society included a large component of middle-class farmers and craftsmen. This was in sharp contrast with European societies in which poor peasants and farm laborers, with very little income beyond subsistence needs, constituted a large fraction of the total population, while at the top of the social pyramid was a small group of large landowners whose expenditure patterns made little contribution to the emergence of a sector producing standardized manufactured goods.

In America, to a degree which is now largely lost from the collective memory of a highly urbanized population, markets in the first half of the nineteenth century were dominated by the tastes and requirements of middle-class rural households.[12] In such households, often living under frontier or at least isolated conditions, there was little concern—indeed there was little opportunity—for ostentation or conspicuous consumption. Out of these social conditions emerged a relative simplicity of taste and a stress upon functionalism in design and structure. Rural isolation also strongly favored reliability of performance and ease of repair in case of breakdown. Such characteristics were particularly important for agricultural machinery and firearms. Simplicity of design and uniformity of components meant that a farmer could repair a broken plough in the field himself without having to depend upon a skilled repairman or distant repair shop. Thus, out of the social and geographic conditions of land-abundant America emerged a set of tastes and preferences highly congenial to

---

[9]Lebergott's estimates of the percentage of the labor force in agriculture are as follows: 1810 – 80.9%; 1820 – 78.8%; 1830 – 68.8%; 1840 – 63.1%; 1850 – 54.8%; 1860 – 52.9%; 1870 – 52.5%; 1880 – 51.3%. Stanley Lebergott, "Labor Force and Employment, 1800–1960," in *Output, Employment and Productivity in the U.S. after 1800*, ed. Dorothy Brady, Studies in Income and Wealth, vol. XXX, National Bureau of Economic Research, New York, 1966, p. 119.

[10]Albert Fishlow, "Comparative Consumption Patterns, the Extent of the Market, and Alternative Development Strategies," in *Micro-Aspects of Development*, ed. Eliezer Ayal, Praeger Publishers, New York, 1973.

[11]"The very fact that food prices should have been relatively lower earlier would have given them more leverage at precisely the right moment—when per capita incomes were smaller and the percentages allocated to food higher. What is significant is that they did not rise rapidly thereafter and impede the extension of the market. American food prices in the 1830s and 1840s were not only lower relative to those of a nonfood composite, but also compared to British foodstuffs. That is, British food prices fell more rapidly from 1831–50 to 1866–90 than did American, and we have already seen that American foods were absolutely cheaper at that later date. Hence, the comparative advantage the United States enjoyed was greater in the earlier period." *Ibid.*, pp. 77–78.

[12]"The share in aggregate expenditures of different groups in the population in the nineteenth century and earlier can be estimated only roughly, for want of statistical information representative of all sections of the country. Even a crude statistical picture, however, can indicate how important rural households were in the market for manufactured goods and services. In the 1830s, about 66 percent of all households were on farms, 23 percent in villages, and 11 percent in cities. On the supposition that farm families on the average were able to spend $100, village families $200, and city families $300 over and above their outlays for housing and fuel, rural families (farm and village) would have accounted for 76 percent of all such expenditures." D. Brady in Davis, Easterlin, and Parker (eds.), *American Economic Growth*, p. 62.

---

a technology capable of producing large quantities of standardized, low-priced goods. These circumstances even left their indelible imprint on the American automobile in the early years of the twentieth century. The Ford Model T was designed in a manner which strongly resembled the horse and buggy, and the primary buyers were farmers for whom a cheap car offered a unique opportunity for overcoming rural isolation.

■ Supply

The point has already been made that in the early nineteenth century America was a large-scale borrower of European technology. But the point also has been made that America's factor proportions, especially its rich abundance of natural resources, differed substantially from Europe's. These two propositions have an interesting implication that provides a useful way of entering into an examination of supply-side influences upon the emergence of the American system of manufactures. For, if factor proportions between the two continents were sufficiently disparate, it would follow that the technology devised to accommodate European needs did not always constitute, in present-day terminology, an "appropriate technology" in America. We now need to consider, therefore, how differences in the supply of available resources pushed Americans in a direction that helps to account for the country's unique technological contributions.

An abundance of natural resources means, in economic terms, that it is rational to employ methods of production which are resource intensive. Much of American inventive activity in the first half of the nineteenth century aimed at substituting abundant natural resources for scarcer labor and capital—although this aim is often not apparent from a present-day examination of the hardware devised for the purpose. America's early world leadership in the development of specialized woodworking machinery—machines for sawing, planing, mortising, tenoning, shaping, and boring—was a consequence of an immense abundance of forest products. Although these machines were wasteful of wood, that was of little consequence in a country where wood was very cheap. The substitution of abundant wood for scarce labor was, in fact, highly rational.[13]

Similarly, the immense available supply of potential farmland was not conducive to inventive activity that would maximize output per acre (as would be sensible in a land-scarce economy like Japan's), but rather to activity that would maximize the amount of land that could be cultivated by a single worker. This is precisely the thrust of nineteenth-century American mechanical innovation in agriculture, based upon animal power which supplied the traction for the steel plough (as well as Jethro Wood's earlier plough which already had replaceable cast-iron parts); the cultivator which replaced the hand-operated hoe in the corn and cotton fields; and the awesome reaper which swept away what once had been a basic constraint upon grain cultivation—fluctuations in labor requirements that reached a sharp peak during the brief harvesting season. Later, output per worker was farther increased by binders and threshing machines and, eventually, by combine harvesters. In corn cultivation, the corn shelter and the corn picker were particularly valuable for their laborsaving characteristics.

In the gunmaking trade, usually regarded as the *locus classicus* of the distinctly American mass-production technology, some of the most original contributions were in the development and elaboration of a set of lathes for shaping the gunstock with a minimum amount of labor. The fact that these laborsaving machines were initially highly wasteful of wood—as compared to the shaping of a gunstock by a Birmingham craftsman—was a trivial consideration in a wood-abundant economy. Blanchard lathes were widely adopted in America.

What is the relevance of these woodworking and agricultural machines to the emergence, by midcentury, of the American system of manufactures? The essential point is that resource abundance provided an incentive in America to explore the possibilities of certain new machine technologies earlier and more deeply than in Europe.[14] The fact that these machines, particularly in their early stages of development, were not only laborsaving but also resource intensive, was not the economic deterrent that it was in Britain or France. Thus, in a wide range of manufacturing activities, Americans had a strong incentive to develop and adopt new technologies which, in effect, traded off abundant natural-resource inputs for labor, as well as machines which were

---

[13]For a more detailed treatment, see Nathan Rosenberg, *Perspectives on Technology*, Cambridge, 1976, chapter 2, "America's Rise to Woodworking Leadership."

[14]The rest of this paragraph draws upon Nathan Rosenberg, "American Technology: Imported or Indigenous?" *American Economic Review Papers and Proceedings*, Feb. 1977.

wasteful of natural resources yet could be constructed more cheaply. But the pressures generated by America's unique resource endowment led to exploratory activities and to eventual learning experiences the outcome of which cannot be adequately summarized merely in terms of factor-saving of factor-using biases. For they also led to new patterns of specialization and division of labor between firms—especially between the producers and users of capital goods—as a result of which the American economy developed a degree of technological dynamism and creativity greater than existed in other industrial economies in the second half of the nineteenth century.

I believe that this technological dynamism was due in large measure to the unique role played by the capital-goods industries in the American industrialization process and the especially favorable conditions under which they operated. For these capital-goods industries—I refer here primarily to those involved in the forming and shaping of metals—became learning centers where metalworkihg skills were acquired and developed, and from which such skills were eventually transferred to the production of a sequence of new standardized products—interchangeable firearms, clocks and watches, agricultural machinery, hardware, sewing machines, typewriters, office machinery, bicycles, automobiles. A key feature of the story of American industrialization is that it involved the application of certain basically similar production techniques to a growing range of manufactures. Moreover, the technological knowledge and competence that gradually accumulated in this sector were directly applicable to generating cost reduction in the production of capital goods themselves. A newly designed turret lathe or universal milling machine, or a new steel alloy permitting a lathe to remove metal at higher speeds—each of these innovations not only resulted in better machines but also reduced the cost of producing the machines in the first place.

Thus, although the initial shift to the capital-using end of the spectrum was generated by the unique pattern of American resource scarcities, it is by no means obvious that the final outcome of this process was in terms of factor biases. For the capital-using path was also a path which, *eventually*, generated a much-increased capacity for capital-saving innovations. The attempt to deal with labor scarcity in a regime of natural-resource abundance pushed us quickly in a direction in which there turned out to be rich inventive possibilities. In turn, the skills acquired in a more capital-abundant society with an effectively organized capital-goods sector provided the basis—in terms of knowledge and engineering skills and expertise—for innovations which were capital saving as well as labor saving. Indeed, most new products, after their technical characteristics became sufficiently stabilized, have passed through such a cost-reducing stage during which capital-goods producers accommodated themselves more efficiently to the large-quantity production of the new product. American industry seems to have particularly excelled at these activities.

Aside from the highly visible major inventions, capital-intensive technologies have routinely offered extensive opportunities for improvements in productivity which seem to have had no equivalent at the labor-intensive end of the spectrum. Knowledge of mechanical engineering, metallurgy, and, perhaps most important of all, the kind of knowledge which comes from day-to-day contact with machine technology, provide innumerable opportunities for small improvements—minor modifications, adaptation to some special-purpose use, design alterations, substitution of a superior or cheaper material—the cumulative effects of which have, historically, been very great.

It is important that these developments be seen in their actual historical sequence. America began the growth of her capital-goods sector not only with a strong preoccupation with standardization and interchangeability, but also with some early experience with techniques as well as a market which readily accepted standardized products. This shaped the eventual outcome of the industrialization process in some decisive ways. The acceptance of standardization and interchangeability vastly simplified the production problems confronting the makers of machinery and provided the technical basis for cost reductions in machinemaking. At the same time it provided the conditions which encouraged the emergence of highly specialized machine producers as well as the transfer of specialized technical skills from one industrial application to another. Indeed, America's most significant contributions to machine-tool design and operation and related processes—profile lathes, turret lathes, milling machines, die-forging techniques, drilling and filing jigs, taps and gauges—were associated with specialized, high-speed machinery devoted to the production of standardized components of complex products.

Thus, as a result of certain initial conditions, America's industrialization in the early nineteenth century proceeded along one out of a variety of possible paths. It was a path dictated by peculiar resource conditions and also a path rich in inventive possibilities. Movement along this path generated a dynamic interaction of forces, a dialectical process in which cause and effect become exceedingly difficult to disentangle. While greater homogeneity of tastes was originally conducive to the introduction of goods produced according to the American system of manufactures, it is also true that, once this technology began to spread, it in turn shaped and influenced tastes in the direction of simplicity and functionality.[15] Furthermore, although American factor endowment pushed in the direction of mechanization, the experience with mechanization *in itself* brought about an improvement in inventive ability and its more rapid diffusion. This was an interactive process, whereby the successful application of mechanical skills in one sector improved the likelihood of their successful application in other sectors.

What, finally, can we add by way of explaining why Britain—which was far more advanced industrially than the United States in the early nineteenth century—did not go further than it did in developing the distinctive features of the American system of manufactures? Much of the answer is already implicit in the preceding discussion of Anglo-American differences in conditions of demand-and-resource endowment. There is, however, an additional factor which deserves mention, namely, the persistence and continuing powerful influence of certain traditional values and attitudes which had their roots in a preindustrial craft society. Such values and attitudes had, of course, played a much more significant role in shaping the outlook of workers and engineers in Britain than in the United States. Strong craft traditions, with their emphasis upon pride in workmanship, individuality, and high standards of product quality, were often inimical to standardization and the alterations in final product design which were essential to low-cost, high-volume production. To some extent the persistence of traditional craft attitudes in Britain reflected the demand-side phenomena to which we have already referred, as well as other market peculiarities. But beyond that there is evidence from many industries—firearms, automobiles, locomotives, a wide range of production machinery, clocks and watches—suggesting a preoccupation with technical perfection beyond what could be justified by narrowly utilitarian considerations.[16] And perhaps this observation suggests what should be an appropriate final comment upon why this particular system first emerged in America. For the American system of manufactures was, above all, a totally unsentimental approach to the productive process in industry, one in which purely commercial considerations prevailed. The British long consoled themselves with the belief that standardization and mass-production techniques inevitably resulted in an inferior product. Unhappily, from their point of view, they persisted in this belief long after it ceased to contain even an element of truth.

---

[15] Dorothy Brady, "Relative Prices in the Nineteenth Century," *Journal of Economic History*, 24 (1964), pp. 147–148.

[16] S.B. Saul, "The Market and the Development of the Mechanical Engineering Industries in Britain, 1860–1914," *Economic History Review*, 41 (1967); S.B. Saul, "The Engineering Industry," chapter 7 in *The Development of British Industry and Foreign Competition, 1875–1914*, ed. D. Aldcroft, University of Toronto Press, Toronto, 1968; S.B. Saul, "The Motor Industry in Britain to 1914," *Business History*, 5 (1962); R.A. Church, "Nineteenth Century Clock Technology in Britain, the U.S., and Switzerland," *Economic History Review,* 49 (1975); and Nathan Rosenberg, introduction to *The American System of Manufactures*, Edinburgh University Press, Edinburgh, 1969.

The System Must Be First

Reading 4

Thomas Hughes

[W]e reflect too little about the influences and patterns of a world organized into great technological systems. Usually we mistakenly associate modern technology not with systems but with such objects as the electric light, radio and television, the airplane, the automobile, the computer, and nuclear missiles. To associate modern technology solely with individual machines and devices is to overlook deeper currents of modern technology that gathered strength and direction during the half-century after Thomas Edison established his invention factory at Menlo Park. Today machines such as the automobile and the airplane are omnipresent. Because they are mechanical and physical, they are not too difficult to comprehend. Machines like these, however, are usually merely components in highly organized and controlled technological systems. Such systems are difficult to comprehend, because they also include complex components, such as people and organizations, and because they often consist of physical components, such as the chemical and electrical, other than the mechanical. Large systems—energy, production, communication, and transportation—compose the essence of modern technology. . . .

. . . In seeking the creators of modern industrial America, we must consider the system builders as well as the independent inventors and the industrial scientists.

Henry Ford's production system remains the best known of the large technological systems maturing in the interwar years. Contemporaries then usually perceived it as a mechanical production system with machine tools and assembly lines. But Ford's system also included blast furnaces to make iron, railroads to convey raw materials, mines from which these came, highly organized factories functioning as if they were a single machine, and highly developed financial, managerial, labor, and sales organizations. . . .

Electrical light-and-power systems, such as those managed and financed by the system-building utility magnate Samuel Insull of Chicago, incorporated not only dynamos, incandescent lamps, and transmission lines, but hydroelectric dams, control or load-dispatching centers, utility companies, consulting-engineering firms, and brokerage houses, as well. When Ford placed a mechanical assembly line in motion, the public was greatly impressed, but electrical systems transmitted their production units too rapidly to perceive: 186,000 miles per second, the speed of light. . . .

Of the American system builders, none took on a more difficult and controversial task than Frederick Winslow Taylor. Ford directed his ordering-and-controlling drive primarily to production machines; Insull focused his on ensuring the large and steady flow of electrical power; Taylor tried to systematize workers as if they were components of machines. Ford's image was of a factory functioning as a machine; Insull envisaged a network or circuit of interacting electrical and organizational components; and Taylor imagined a machine in which the mechanical and human parts were virtually indistinguishable. Idealistic, even eccentric, in his commitment to the proposition that efficiency would benefit all Americans, Taylor proved naïve in his judgments about complex human values and motives. In the history of Taylorism we find an early and highly significant case of people reacting against the system builders and their production systems, a reaction widespread today among those who fear being co-opted by "the system." . . .

Taylor was not the first to advocate a so-called scientific approach to management, but the enthusiasm and dedication, bordering on obsession, with which he gave himself to spreading his views on management, his forceful personality, and his highly unusual and erratic career filled with failure as well as success have left a strong, indelible impression on his contemporaries and succeeding generations. More than a half-century after his death, many persons in Europe, the Soviet Union, and the United States continue to label scientific management "Taylorism." Labor-union leaders and radicals then and now find Taylor convenient to attack as a symbol of a despised system of labor organization and control. In the early decades of this century, Europeans and Russians adopted "Taylorism" as the catchword for the much-admired and -imitated American system of industrial management and mass production. The publication in 1911 of Taylor's *Principles of Scientific Management* remains a landmark in the history of management-labor relations. Within two years of publication, it had been translated into French, German, Dutch, Swedish, Russian, Italian, Spanish, and Japanese. In his novel *The Big Money* (1936), John Dos Passos gave a sketch of Taylor, along with ones of Edison, Ford, Insull, and a few others, because he believed that they expressed the spirit of their era. Dos Passos noted that Taylor never smoked or drank tea, coffee, or liquor, but found comparable stimulation in solving problems of efficiency and production. For him, production was the end-all, whether it be armor plate for battleships, or needles, ball bearings or lightning rods.

Taylor's fundamental concept and guiding principle was to design a system of production involving both men and machines that would be as efficient as a well-designed, well-oiled machine. He said, "in the past, the man has been first; in the future the system must be first," a remark that did not sit well then with workers and their trade-union leaders and that today still rankles those who feel oppressed by technology. He asked managers to do for the production system as a whole what inventors and engineers had done in the nineteenth century for machines and processes. Highly efficient machines required highly efficient functionally related labor. When several Taylor disciples, including later U.S. Supreme Court Justice Louis D. Brandeis, sought a name for Taylor's management system, they considered "Functional Management," before deciding on "Scientific Management." Taylor and his followers unfeelingly compared an inefficient worker to a poorly designed machine member.

Taylor developed his principles of management during his work as a machinist and then as a foreman in the Midvale Steel Company of Philadelphia. . . .

Worker soldiering, variously called "stalling," "quota restriction," "goldbricking" "by Americans, *Bremsen* by Germans, and "hanging it out" or "Ca'canny" by the English and Scots, greatly offended Taylor's sense of efficiency. Having concluded that workers, especially the skilled machinists, were the major industrially inefficient enclave remaining after the great wave of nineteenth-century mechanization, Taylor proposed to eliminate "soldiering." He later wrote that "the greater part of systematic soldiering . . . is done by the men with the deliberate object of keeping their employers ignorant of how fast work can be done." The machinists at Midvale, for example, were on a "piecework" schedule, so they were determined that the owners not learn that more pieces could be turned out per hour and therefore demand an increase in the number of pieces required. They did not trust the owners to maintain the piece rate and allow the workers, if they exerted themselves, to take home more pay. The workers believed that the increased effort would become the norm for the owners. We can only conjecture about the natural rhythm and reasonableness of the pace that the workers maintained over the long duration; Taylor believed that they were soldiering. Nevertheless, he also showed that he was determined that the diligent worker be rewarded with a share of the income from more efficient and increased production. To his consternation, he later found that management and the owners also soldiered when it came time to share the increased income. Taylor was no close student of human nature; his approach was, as he described it, scientific.

After being put in charge of the machinists working at the lathes, Taylor set out to end soldiering among them. His friends began to fear for his safety. As Taylor recalled, the men came to him and said, "Now, Fred, you are not going to be a damn piecework hog, are you?" To which he replied, "If you fellows mean you are afraid I am going to try to get a larger output from these lathes," then "Yes; I do propose to get more work out." The piecework fight was on, lasting for three years at Midvale. Friends begged Taylor to stop walking home alone late at night through deserted streets, but he said that they could shoot and be damned and that, if attacked, he would not stick to the rules, but resort to biting, gouging, and brickbats. At congressional hearings in 1912—about thirty years later—he insisted:

> I want to call your attention, gentlemen, to the bitterness that was stirred up in this fight before the men finally gave in, to the meaness of it. . . . I did not have any bitterness against any

particular man or men. Practically all of those men were my friends, and many of them are still my friends. . . . My sympathies were with workmen, and my duty lay to the people by whom I was employed.

In his search for the one best way of working, of deciding how and how fast a lathe operator should work, he used a method that he considered scientific. He believed values and opinions of neither workers nor managers influenced his objective, scientific approach. Beginning in 1882, first he, then an assistant began using a stopwatch to do time studies of worker's motions. Timing was not a new practice, but Taylor did not simply time the way the men worked: he broke down complex sequences of motions into what he believed to be the elementary ones and then timed these as performed by workers whom he considered efficient in their movements. Having done this analysis, he synthesized the efficiently executed component motions into a new set of complex sequences that he insisted must become the norm. He added time for unavoidable delays, minor accidents, inexperience, and rest. The result was a detailed set of instructions for the worker and a determination of time required for the work to be efficiently performed. This determined the piecework rate; bonuses were to be paid for faster work, penalties for slower. He thus denied the individual worker the freedom to use his body and his tools as he chose.

Taylor stressed that the time studies, with their accompanying analysis and synthesis, did not alone constitute scientific management. He realized and insisted that, for the work to be efficiently performed, the conditions of work had to be reorganized. He called for better-designed tools and became known for his near-fixation about the design of shovels. He ordered the planning and careful management of materials handling so that workers would have the materials at hand where and when needed. Often, he found, men and machines stood idle because of bottlenecks in complex manufacturing processes. Taylor even attended to lighting, heating, and toilet facilities. Seeing inanimate machines and men together as a single machine, he also looked for ways in which the inanimate ones failed. Believing that machine tools could also be driven faster, he invented a new chromium-tungsten steel for cutting tools that greatly increased their speed. As we would expect, he did not leave decisions about even the cutting speed of the machine tool or the depth of the cut to the subjective judgment of the machinist. In his book *On the Art of Cutting Metals* (New York, 1907), he described his thousands of experiments that extended over twenty-six years.

As a system builder seeking control and order, Taylor was not content to redesign machines, men, and their relationship; he was set upon the reorganization of the entire workplace or factory as a machine for production. Stimulated by his example, individuals with special education, training, and skill contributed to the establishment of "the new factory system." To understand this achievement, we need to consider the way in which the work process was carried out in many machine shops, engineering works, and factories before Taylor's reforms. After the concern received an order, copies specifying the product and quantity to be made were sent to the foremen. They carried most of the responsibility for the production process. Once draftsmen had prepared detailed drawings, foremen in the machine shop, the foundry, pattern-making shop, and forge determined the various component parts needed, ordered the raw materials, and wrote out job cards for the machinists. The machinists then collected drawings, raw materials, and tools, and planned the way in which the job for a particular component part would be done. When the machinists had completed the particular job, they reported to the foreman for another. The foremen had overall supervision, but there was little scheduling and, therefore, little planned coordination of the various jobs. Components sometimes reached the assembly point, or erecting shop, haphazardly. Because of lack of planning, scheduling, and close monitoring of the progress of work, raw materials were often not on hand. How the workmen might then use their time is not clear, but proponents of Taylorism leave the impression that they were idle.

Taylor found the disorder and lack of control unbearably inefficient and declared war on traditional methods responsible for these. His reform specified that an engineering division take away from the foremen overall responsibility for the preparation of drawings, the specification of components, and the ordering of raw materials. Upwardly mobile young graduates from the rising engineering schools were soon displacing their "fathers," the foremen. The planning department in the engineering division coordinated deliveries of materials, and the sequence in which component parts would be made. The planning department prepared detailed instructions about which machines would be used, the way in which machinists, pattern makers, and other workmen would make each part, and how long the job should take. Careful records were kept of the progress being made in the manufacture of each part, including materials used and time consumed. Unskilled workers moved materials and parts around shops so that they would be on hand where and when needed. By an elaborate set of instruction cards and reports, the planning department had an overall picture of the flow of parts throughout the shops, a flow that prevented the congestion of the work at particular machines and the idleness of other machines and workmen. The reports of worker time and materials consumed greatly facilitated cost accounting.

The complexity and holism of Taylor's approach was often ignored because of the widespread publicity given to some of his simplest and most easily reported and understood successes. Taylor often referred to the "story of Schmidt," who worked with the pig-iron gang at the Bethlehem Steel Corporation in Pennsylvania. When Taylor and his associates came to Bethlehem in 1897 to introduce their management techniques and piecework, they found the pig-iron gang moving on the average about twelve and a half tons per day. Each man had repeatedly to lift ninety-two pounds of iron and carry it up an inclined plant onto a railroad car. After careful inquiry into the character, habits, and ambition of each of the gang of seventy-five men, Taylor singled out a "little Pennsylvanian Dutchman who had been observed to trot backhome for a mile or so after his work in the evening about as fresh as he was when he came trotting down to work in the morning." After work he was building a little house for himself on a small plot of ground he had "succeeded" in buying. Taylor also found out that the "Dutchman" Henry Noll, whom Taylor identified as Schmidt, was exceedingly "close," or one who placed "very high value on the dollar." The Taylorites had found their man.

Taylor recalled the way he and Schmidt talked, a story that tells us more of Taylor's attitudes than of what actually transpired:

> "Schmidt, are you a high-priced man? . . . What I want to find out is whether you want to earn $1.85 a day or whether you are satisfied with $1.15, just the same as all those cheap fellows are getting?"
>
> "Did I vant $1.85 a day? Vas dot a high-priced man? Vell, yes, I vas a high-priced man."
>
> ". . . Well, if you are a high-priced man, you will load that pig iron on that car tomorrow for $1.85. You will do exactly as this man tells you tomorrow, from morning till night. When he tells you to pick up a pig, and walk, you pick it up and you walk, and when he tells you to sit down and rest, you sit down. . . . And what's more, no back talk."

Taylor found it prudent to add:

> This seems to be rather rough talk. And indeed it would be if applied to an educated mechanic or even intelligent laborer. With a man of the mentally sluggish type of Schmidt it is appropriate and not unkind, since it is effective in fixing his attention on high wages which he wants. . . .

Perhaps Taylor, the upper-middle-class Philadelphian, forgot that the Pennsylvania Dutchman was not so mentally sluggish that he could not save for land and build a house. Schmidt moved the forty-seven tons of pig that the Taylorites had decided should be the norm, instead of the former twelve and a half tons, and soon all the gang was moving the same and receiving sixty percent more pay than other workmen around them. We are not told whether Schmidt was still able to trot home and work on his house.

Numerous other examples of Taylor's methods increasing worker output and production abound, but there is also abundant evidence of failures. Ultimately his efforts at Bethlehem Steel exhausted him, and the head of the company summarily dismissed him. Taylor had come to the steel company with the full support of Joseph Wharton, a wealthy Philadelphian into whose hands the company had passed. Wharton wanted a piecework system installed in the six-thousand-man enterprise. Taylor warned that his system would be strongly opposed by all of the workmen, most of the foremen, and even a majority of the superintendents. Bold and determined, he forged relentlessly ahead, introducing a planning department and new administrative roles for the foremen. Instructions for routine were codified with time cards, work sheets, order slips, and so on. As worker resistance stiffened over several years, Taylor became rigid, even arbitrary, in dealing with labor and management. His achievements were impressive, but "as time went on, he exhibited a fighting spirit of an intensity almost pathological," an admirer wrote. Taylor's communications to the Bethlehem president were tactless and peremptory (he believed Wharton would shelter him). He complained of poor health and nervous strain. He thought that some of the major stockholders opposed him because he was cutting the labor force, and they were losing rents on the workers' houses. The curt note dismissing him came in April 1901.

Many workers were unwilling, especially the skilled ones, to give control of their bodies and their tools to the scientific managers, or, in short, to become components in a well-planned system. An increase in pay often did not compensate for their feeling of loss of autonomy. Taylor's scientific analysis did not take into account worker independence and pride in artful craftsmanship—even artful soldiering. Perhaps this was because Taylor, despite his years of experience on the shop floor, did not come from a blue-collar worker culture.

Samuel Gompers, a labor leader, said of Taylorism and similar philosophies of management:

> So, there you are, wage-workers in general, mere machines—considered industrially, of course. Hence, why should you not be standardized and your motion-power brought up to the highest possible perfection in all respects, including speeds? Not only your length, breadth, and thickness as a machine, but your grade of hardness, malleability, tractability, and general serviceability, can be ascertained, registered, and then employed as desirable. Science would thus get the most out of you before you are sent to the junkpile.

One of the most publicized setbacks for Taylorism took place at the Watertown Arsenal when Carl G. Barth, a prominent Taylor follower and a consultant on scientific management, tried to introduce the Taylor system. Serious trouble started in the foundry when one of Barth's associates began stopwatch-timing the men's work procedures. The skilled workers in the shop discovered that the man carrying out the study knew little about foundry practice. The foundrymen secretly made their own time study of the same work process and complained that the time specified by the "expert" was uninformed and represented an unrealistic speedup. The Watertown project was also flawed because Taylor's practice was to reorganize and standardize a shop before doing time-and-motion studies, and this had not been carried out at Watertown Arsenal. On the evening after the initiation of the stopwatch studies, the workers met informally and in a petition to the commanding officer of the arsenal they stated:

> The very unsatisfactory conditions which have prevailed in the foundry among the molders for the past week or more reached an acute stage this afternoon when a man was seen to use a stop watch on one of the molders. This we believe to be the limit of our endurance. It is humiliating to us, who have always tried to give to the government the best that was in us. This method is un-American in principle, and we most respectfully request that you have it discontinued at once.

When stopwatch timing continued, the molders walked out on 11 August 1911.

Promised an investigation of the "unsatisfactory conditions," the molders returned to work after a week, but the publicity given a strike against the U.S. government intensified, fermenting union opposition to scientific management, specifically Taylorism, at Watertown and at another U.S. arsenal, at Rock, Illinois. August brought the formation of a special congressional committee of three to investigate scientific management in government establishments. The committee took extensive testimony from Taylor, among others. He became so exercised by hostile questions that his remarks had to be removed from the record. The report of the committee did not immediately call for any legislation. In 1914, however, Congress attached to appropriations bills the proviso that no time studies or related incentive payments should be carried out in government establishments, a prohibition that survived for over thirty years. Yet Taylorism involved, as we have seen, more than time studies and incentive payments, so work processes in government establishments continued to be systematically studied, analyzed, and changed in ways believed by management experts to be scientific.

Worn down by the never-ending opposition and conflict, Taylor moved in 1902 to a handsome house in me Chestnut Hill area of Philadelphia. He no longer accepted employment or consulting fees but announced that he was ready to advise freely those interested in Taylorism. . . .

Free from the confrontations in the workplace, Taylor dedicated himself to showing that his philosophy of management would ultimately promote harmony between management and labor. He argued that increasing production would increase wages and raise the national standard of living. His principles of scientific management struck responsive chords in a nation intent on ensuring economic democracy, or mass consumption, through mass production and also on conserving its natural resources. Taylor wrote that maximum prosperity could exist only as a result of maximum productivity. He believed that the elimination of wasted time and energy among workers would do more than socialism to diminish poverty and alleviate suffering.

Because of his firm belief that his method was objective, or scientific, he never fully comprehended the hostile opposition of aggressive, collective-bargaining labor-union leaders. He found the unions mostly standing "for war, for enmity," in contrast to scientific management, which stood for "peace and friendship." Nor could he countenance unenlightened and "hoggish" employers who either found his approach and his college-educated young followers unrealistic or were unwilling to share wholeheartedly with the workers the increased profits arising from scientific management. He considered the National Manufacturers Association a "fighting association," so he urged his friends in scientific management to cut all connections with it and its aggressive attitudes toward labor unions. Firmly persuaded that conflict and interest-group confrontations were unnatural, he awaited, not too patiently, the day when management and labor would realize, as he, that where the goal was increased productivity there were discoverable and applicable scientific laws governing work and workplace. Scientific managers were the experts who would apply the laws. He wrote:

> I cannot agree with you that there is a conflict in the interests of capital and labor. I firmly believe that their interests are strictly mutual, and that it is practicable to settle by careful scientific investigation the proper award that labor should receive for the work it renders.

Their interest was not only a mutual but a national one—production and democracy, production and Democracy. Taylor's times were not ones of affluence for workers, so his means to the end of mass production, thereby raising living standards of the glasses, seemed in accord with democratic principle. Within a few years, Vladimir Lenin argued that Taylor principles accorded with socialism, as well.

Taylor became nationally known when [Louis] Brandeis, the Boston "people's lawyer," argued in 1911 that scientific management especially Taylorism, could save the nation's railroads so much money that the increased rates that the railroads were requesting from the Interstate Commerce Commission would not be needed. Since the rate hearings were well publicized, writers from newspapers and magazines descended on Taylor to find out about his system and then, at his suggestion, visited Philadelphia plants to see firsthand Taylorism in practice. The favorable publicity induced Taylor to write that "the interest now taken in scientific management is almost comparable to that which was aroused in the conservation of our natural resources by Roosevelt."

Taylor rightly associated his scientific management with the broader conservation movement that had attracted national interest and support during Theodore Roosevelt's terms as president, 1901–08. This progressive program for conservation focused on the preservation and efficient utilization of lands and resources. Like scientific management, it advocated that decisions about conservation be made scientifically by experts. Like Taylor, the progressive conservationists did not countenance as inevitable conflict of interests among ranchers, farmers, lumbermen, utilities, manufacturers, and others. To the contrary, they believed that such conflict was regressive, that it must be displaced by a scientific approach expected to bring harmonious and rational compromises in the general interest. This approach expressed a technological spirit spread by engineers, professional managers, and appliers of science, a belief that there was one best way. College-educated foresters, hydraulic engineers, agronomists should be, the progressives argued, the decision makers about resources; professional managers about the workplace.

Taylor and the growing number of his followers wrote books, published articles, gave lectures, and acted as consultants. He authorized C. G. Barth, H. K. Hathaway, Morris L. Cooke, and Henry L. Gantt to teach his system of management: "All others were operating on their own." Frank Gilbreth, among those who operated "on their own," became well known for his *A Primer of Scientific Management* (1914) and for the use he and his wife, Lillian Gilbreth, made of the motion-picture camera to prepare time-and-motion studies. Her contribution to scientific management has yet to be generally acknowledged. She, not her husband, had a Ph.D. degree in psychology (Brown University, 1915). Perhaps because of her study of psychology, she sensitively took into account complex worker characteristics. The Gilbreths' articles on scientific management show the influence of her concern that the worker should not be seen simply as a component in a Taylor system. After her husband's death, she continued her consulting work and served as a professor of industrial management at Purdue University.

Mass Production

Reading 5

David A. Hounshell

> Mass production became the Great American art.
> —Paul Mazur, *American Prosperity (1920)*

Since the 1920s the term "mass production" has become so deeply ingrained in our vocabulary and our thought that we seldom stop to ask how it arose and what lay behind its appearance. The purpose of this brief essay is to provide an overview of the development of mass production in America as a means of getting at these questions. In the first half of the nineteenth century, manufacturing in the United States developed along such distinct lines that by the 1850s English observers came to speak of an "American system" of manufactures. Subsequently, the American system grew and changed in character so much that by the 1920s, the United States possessed the most prolific production technology the world has ever known. This was "mass production."

In 1925, the American editor of the *Encyclopaedia Britannica* wrote to Henry Ford asking him to submit an article on "Mass Production" for the three-volume supplement to the *Britannica*, the so-called 13th Edition. Apparently Ford's office, if not Ford himself, responded favorably and promptly set Ford's spokesman, William J. Cameron, to work on the article. Cameron consulted the company's chief production planner about how the "general reader" might comprehend the principles of mass production. When Cameron completed the article, he placed Henry Ford's name beneath it and sent it to the *Britannica's* New York office.

Although Cameron would later say that he "should be very much surprised to learn that [Henry Ford] read it," this article played a fundamental role in giving the phrase "mass production" a place in the English vocabulary. Even before the article appeared in the Britannica, the *New York Times* published it as a full-page, feature article in a Sunday edition. Under the banner, "Henry Ford Expounds Mass Production: Calls It the Focusing of the Principles of Power, Economy, Continuity and Speed," the article attracted the attention of a wide segment of the American population, especially since it also went through the wire service. While one can certainly wonder what led the *Britannica* editor to choose the term, mass production, there is little doubt that the ghost-written Ford article led to the widespread use of the term and its identification with the assembly line manufacturing techniques that were the hallmark of automobile production. Immediately, the article proved interesting enough to provoke a *Times* editorial. The term, which had not commonly appeared in reference works such as the *Reader's Guide to Periodic Literature* prior to the *Britannica Times* article, soon passed into general use in both popular and scholarly literature. After the appearance of "Mass Production," the previously popular expression, "Fordism," soon disappeared. The Ford article endowed mass production with a certain universality despite its ambiguity and its status as poor grammar.

Much more important than the story of how mass production entered the English vocabulary are the developments that lay behind the manufacturing system described in the article. Commenting in 1940 on Henry Ford and the *Britannica* article in his *Engines of Democracy*, Roger Burlingame raised the essential questions:

> With [Ford's] great one-man show moving toward a dictatorship of which any totalitarian leader might well be proud he was ready for what he calls [Mass] Production. [Mass] Production, Ford believes, had never existed in the world before. With the magnificent contempt of men

"Mass Production" by David A. Hounshell as appeared in "Mass Production in American History, 1800–1932" from Polhem 2 *(1984): 1–28.*
© David A. Hounshell. Reprinted by permission of the author.

immune to history, he disregards all predecessors: Whitney, Evans, Colt, Singer, McCormick, the whole chain of patient, laborious workers who wrought his assembly lines and all the ramifications of his processes out of the void of handicrafts. In a colossal blurb printed in the *Encyclopaedia Britannica* under the guise of an article on mass production, he writes: "In origin, mass production is American and recent; its notable appearance falls within the first decade of the 20th century," and devotes the remainder of the article and two full pages of half-tone plates to the Ford factory.

Burlingame was obviously contemptuous of the claims that mass production was a creation of the Ford Motor Company. Eli Whitney, Oliver Evans, Samuel Colt, Isaac Singer, and Cyrus McCormick, among others, he implied, provided essential building blocks for development at Ford. Burlingame was even more pointed when he later asked rhetorically, "What are those production methods in use today in every large automobile plant with scarcely any variation? They are simply the methods of Eli Whitney and Samuel Colt, improved, coordinated and applied with intelligent economy—economy in time, space, men, motion, money and material."

Since the establishment of the history of technology as an academic discipline in the United States, the assertions contained both in Ford's encyclopaedia article and in Burlingame's popular work have come under close study by a number of investigators. Indeed, the so-called "American system of manufactures," which describes the methods of Whitney, Colt, and the rest, has become one of the most productive areas of American scholarship in the history of technology, and there now exists a rich body of literature on this historical phenomenon. Portions of that new scholarship . . . indicate that the Ford article came much nearer the truth than did Burlingame and his followers. "[I]n origin," as the Ford piece suggested, "mass production is American and recent"—what Whitney et al., did in the nineteenth century was not true mass production. . . .[M]ass production differed in kind as well as in scale from the techniques referred to in the antebellum period as the American system of manufactures. This can be seen most clearly by first considering the American system itself.

Two decades of research on this topic have yielded a number of conclusions, particularly concerning a basic aspect of modern manufacturing, the interchangeability of parts. The symbolic kingpin of interchangeable parts production fell in 1960 when Robert S. Woodbury published his essay, "The Legend of Eli Whitney," in the first volume of *Technology and Culture*. Woodbury convincingly argued that the parts of Whitney's guns were not in fact constructed with interchangeable parts. In 1966, the artifactual research of Edwin A. Battison solidly confirmed Woodbury's more traditional, document-based research findings. Eugene S. Ferguson later wrote of Woodbury's pioneering article, "Except for Whitney's ability to sell an undeveloped idea, little remains of his title as father of mass production."

With Eli Whitney reinterpreted as a *promoter* rather than as a *pioneer* of machine-made interchangeable parts manufacture, it remained for Merritt Roe Smith to identify conclusively the personnel and the circumstances of this fundamental step in the development of mass production. Smith demonstrated that the United States Ordinance Department was the prime mover in bringing about machine-made interchangeable parts production of small arms. The national armory at Springfield, Massachusetts, played a major role in this process, especially as it tried to coordinate its operations with those of its sister armory at Harpers Ferry and John Hall's experimental rifle factory, also located in Harpers Ferry. While these federally owned arms plants occupied a central place in its efforts, the Ordnance Department also used contracts with private arms makers to further its aims. By specifying interchangeability in its contracts and by giving contractors access to techniques used in the national armories, the Ordinance Department contributed significantly to the growing sophistication of metalworking and woodworking (in the case of gunstock production) in the United States by the 1850s. British observers found these techniques sufficiently different from their own and alluded to them in expressions such as the "American system," the "American plan," and the "American principle."

Although British visitors to the United States in the 1850s, especially Joseph Whitworth and John Anderson, were impressed with every aspect of American manufacturing, small arms production received their most careful and detailed analysis. Certainly this was Anderson's job, for he had been sent to the United States to find out everything he could about small arms production and to purchase arms-making machinery for the Enfield Arsenal. In his report, Anderson indicated that the federal armory at Springfield had indeed achieved what the Ordnance Department had sought since its inception: true interchangeability of parts. Anderson and his committee went into Springfield's arsenal and randomly selected ten muskets, each made in a different year from 1844 to 1853. A workman then disassembled these muskets, and their parts were mixed together. According to Anderson, the committee then "requested the workman, whose duty it is to 'assemble' the arms, to put them together, which he did—the Committee handing him the parts, taken at hazard—with the use of a turn-screw only, and as quickly as though they had been English muskets whose parts had carefully been kept separate.

What Anderson was not likely to have known was the extraordinary sum of money that the Ordnance Department had expended over a forty or fifty year period, "[i]n order," as an Ordnance Officer wrote in 1819, "to attain this grand object of uniformity of parts." Nor was Anderson necessarily aware that the unit cost of Springfield small arms with interchangeable parts almost certainly was significantly higher than arms produced by more traditional methods. He should, however, have been aware that the Ordnance Department could annually turn out only a relatively small number of Springfield arms manufactured with interchangeable parts. Despite the high costs and limited output, Anderson pointed out that the special techniques used in the Springfield Armory as well as in some private armories could be applied almost universally in metalworking and woodworking establishments. In fact, by the time Anderson reached this conclusion, the application of those techniques in other industries was already under way.

The new manufacturing technology spread first to the production of a new consumer item, the sewing machine, and eventually it diffused into other areas, including consumer durables such as typewriters, bicycles, and eventually automobiles. Nathan Rosenberg has provided economic and technological historians with an excellent analysis of a major way in which this diffusion occurred. Rosenberg identified the American machine tool industry, which grew out of the small arms industry (notably the Colt armory and the firm of Robbins & Lawrence in Windsor, Vermont, and Hartford, Connecticut) as the key agent for introducing arms-making technology into the sewing machine industry, the bicycle industry, and the automobile industry. The makers of machine tools worked with manufacturers in various industries as they encountered and overcame production problems relating to the cutting, planing, boring, and shaping of metal parts. As each problem was solved, new knowledge went back into the machine tool firms, which then could be used for solving production problems in other industries. Rosenberg called this phenomenon "technological convergence." In many industries that worked with metal, the final products were vastly different in terms of the kinds of markets in which they were sold—the Springfield Armory, for example, "sold" its products to a single customer, the government, while sewing machines producers faced a widely scattered group of individual consumers. Nevertheless, these products had *technological* things in common because their manufacture depended upon similar metalworking technique. These common needs "converged" at the point where the machine tool industry interacted with the firms that bought its machine tools.

Although he did not emphasize the point, Rosenberg recognized that individual mechanics played an equally important role in diffusing know-how as they moved from the firearms industry to sewing machine manufacture to bicycle production and even to automobile manufacture. Examples of such mechanics abound. Henry M. Leland is an obvious example: he worked at Springfield Armory, carried this knowledge to Brown & Sharpe Manufacturing Company when it was making both machine tools and Wilcox & Gibbs sewing machines, next created the Cadillac Motor Car Company and finally the Lincoln Motor Company.

But the process of diffusion was neither as smooth nor as simple as Rosenberg and others would have it. New research suggests that the factories of two of the giants of nineteenth century manufacturing, the Singer Manufacturing Company and the McCormick Harvesting Machine Company, were continually beset with production problems. Previously, many historians attributed the success of these two companies to their advanced production technology. But it now appears that a superior marketing strategy (including advertising and sales techniques and policies) proved to be the decisive factor. . . .

Joseph Woodworth, author of *American Tool Making and Interchangeable Manufacturing,* argued that the "manufacture of the bicycle . . . brought out the capabilities of the American mechanic as nothing else had ever done. It demonstrated to the world that he and his kind were capable of designing and making special machinery, tools, fixtures, and devices for economic manufacturing in a manner truly marvellous; and has led to the installation of the interchangeable system of manufacturing in a thousand and one shops when it was formerly thought to be impractical." Clearly the bicycle industry as a staging ground for the diffusion of armory practice cannot be overemphasized. Rosenberg's idea that the machine tool industry played a leading role in this diffusion applies even more clearly to the bicycle than to the sewing machine. The bicycle boom of the 1890s kept the machine tool industry in relatively good health during the serious depression that began in 1893, and it was accompanied by changes in production techniques.

Entirely new developments occurred in bicycle production—sheet metal stamping and electric resistance welding techniques. These new techniques rivalled in importance the diffusion of older metalworking technologies. During the 1890s, bicycle makers located principally but not exclusively in areas west of New England began to manufacture bicycles with many components (pedals, crank hangers, steering heads, joints, forks, hubs, etc.) made from sheet steel. Punch pressing or stamping operations were combined with the recent invention of electric resistance welding to produce parts at significantly lower costs. This technology would become fundamental to the automobile industry. . . .

Bicycle makers such as [Albert A.] Pope who used traditional armory-type production techniques looked with disdain at those who manufactured bicycles with parts made by the new techniques in pressing and stamping steel. An executive at the Columbia works [Pope's factory] called them cheap and nasty. Despite such views, the one manufacturer that outstripped Pope's production at the peak of the bicycle boom was the Western Wheel Works of Chicago, which made a "first class bicycle out of pressed steel hubs, steering head, sprocket, frame joints, crank hanger, fork, seat, handlebar, and various brackets. Although not quite as expensive as the Columbia, the Western Wheel bicycle ranked high in the top price category among some 200 to 300 manufacturers. Production of this bicycle reached 70,000 in 1896, an output that was significantly less than that of the Ford Model T in 1912, the last full year of its pre-assembly line manufacture.

. . . In terms of production, it is only with the rise of the Ford Motor Company and its Model T that there clearly appears an approach to manufacture capable of handling an output of multi-component durables ranging into the millions each year.

Moreover, the rise of Ford marks an entirely new epoch in the manufacture of consumer durables in America. The Ford enterprise may well have been more responsible for the rise of "mass production," particularly for the attachment of the noun "mass" to the expression, than we have realized. Unlike Singer, McCormick, and Pope, Ford sought to manufacture the lowest priced automobile and to use continuing price reductions to produce ever greater demand. Ford designed the Model T to be a "car for the masses." Prior to the era of the Model T, the word "masses" had carried a largely negative connotation, but with such a clearly stated goal and his company's ability to achieve it, Ford recognized "the masses" as a legitimate and seemingly unlimited market for the most sophisticated consumer durable product of the early twentieth century. Whether Henry Ford envisioned "the masses" as the populace or 'lower orders' of late nineteenth parlance or merely as a large number of potential customers hardly matters, for the results were the same. Peter Drucker long ago maintained that Ford's work demonstrated for the first time that maximum profit could be achieved by maximizing production while minimizing cost. He added that "the essence of the mass-production process is the reversal of the conditions from which the theory of monopoly was deduced. The new assumptions constitute a veritable economic revolution." For Drucker, mass production was as much an economic doctrine as an approach to manufacture. For this reason if for no other, the work of the national armories, Singer, McCormick, Pope et al., differed substantially from Ford's. But Ford was able to initiate this new "economic revolution" because of advances in production technology, especially the assembly line.

Before the adoption of the revolutionary assembly line in 1913, Ford's production engineers had synthesized the two different approaches to manufacture that had prevailed in the bicycle era. On the other hand, Ford adopted the techniques of armory practice. All of the company's earliest employees recalled how ardently Henry Ford had supported efforts to improve precision in machining. Although he knew little about jig, fixture, and gauge techniques, Ford nevertheless became a champion of interchangeability within the Ford Motor Company, and he hired mechanics who knew what was required to achieve that goal. Certainly by 1913, most of the problems of interchangeable parts manufacture had been solved at Ford. In addition to armory practice, Ford adopted sheet steel punch and press work in an important way. Initially he contracted for stamping work with the John R. Keim Company in Buffalo, New York, which had been a major supplier of bicycle components. Soon after opening his new Highland Park factory in Detroit, however, Ford purchased the Keim plant and promptly moved its presses and other machines to the new factory. More and mote Model T components were stamped out of sheet steel rather than being fabricated with traditional machining methods. Together, armory practice and sheet steel work equipped Ford with the capability to turn out virtually unlimited numbers of components. It remained for the assembly line to eliminate the remaining bottleneck—how to put these parts together.

The advent of line assembly at Ford Motor Company in 1913 is one of the most confused episodes in American history. . . . First, the assembly line, once it was first tried on April 1, 1913, came swiftly and with great force. Within eighteen months of the first experiments with moving line assembly, assembly lines were used in almost all sub-assemblies and in the most symbolic mass production operation of all, the final chassis assembly. Ford engineers witnessed productivity gains ranging from fifty percent to as much as ten times the output of static assembly methods. Allan Nevins quite correctly called the moving assembly line "a lever to move the world."

Secondly, there can be little doubt that Ford engineers received their inspiration for the moving assembly line from outside the metalworking industries. Henry Ford himself claimed that the idea derived from the "disassembly lines" of meatpackers in Chicago and Cincinnati. William Klann, a Ford deputy who was deeply involved in the innovation, agreed but noted that an equally important source of inspiration was flour milling technology as practiced in Minnesota. . . . Although there may have been a clear connection in the minds of Ford engineers between "flow production" and the moving assembly line, there is little justification for saying that the assembly line came directly from flour milling. The materials and processes of both were simply too different to support such a view.

The origins of the Ford assembly line are less important than its effect. While providing a clear solution to the problems of assembly, the line brought with it serious labor problems. Already, Ford's highly mechanized and subdivided manufacturing operations imposed severe demands on labor. Even more than previous manufacturing technologies, the assembly line implied that men, too, could be mechanized. Consequently during 1913 the Ford company saw its annual labor turnover soar to 380 percent and even higher. Henry Ford moved swiftly to stem this inherently inefficient turn-over rate. On January 5, 1914, he instituted what became known as the "five-dollar day." Although some historians have argued that this was a wage system that more than doubled the wages of "acceptable" workers, most recently the five-dollar day has been interpreted as a profit sharing plan whereby Ford shared excess profits with employees who were judged to be fit to handle such profits. In any case, the five-dollar day effectively doubled the earnings of Ford workers, and provided a tremendous incentive for workers to stay "on the line." With highly mechanized production, moving line assembly, high wages, and low prices on products, "Fordism" was born.

During the years between the birth of "Fordism" and the wide-spread appearance of the term, "mass production," the Ford Motor Company expanded its annual output of Model T's from 300,000 in 1914 to more than 2,000,000 in 1923. In an era when most prices were rising, those of the Model T dropped significantly—about sixty percent in current dollars. Throughout the Model T's life, Henry Ford opened his factories to technical journalists to write articles, series of articles, and books on the secrets of production at Ford Motor Company. Soon after the appearance of the first articles on the Ford assembly lines, other automobile companies began putting their cars together "on the line." Manufacturers of other consumer durables also followed suit. Ford's five-dollar day forced automakers in the Detroit vicinity to increase their wage scales. Because Ford secured more than fifty percent of the American automobile market by 1921, his actions had a notable impact on American industry.

Ford's work and the emulation of it on the part of other manufacturers led to the establishment of what could be called an "ethos of mass production" in America. The creation of this ethos marks a significant moment in the development of mass production and consumption in America. Certain segments of American society looked at Ford's and the entire automobile industry's ability to produce large quantities of goods at surprisingly low costs. When they did so, they wondered why, for example, housing, furniture, and even agriculture could not be approached in precisely the same manner in which Ford approached the automobile.

Consequently, during the years that the Model T was in production, movements arose within each of these industries to introduce mass production methods. In housing, an industry always looked upon as one of the most staid and pre-industrial of all, prefabrication efforts reached heights not achieved by the pioneers of prefabrication. Foster Gunnison, for example, strove to become the "Henry Ford of housing" by establishing a factory to turn out houses on a moving assembly line, and Gunnison was only one among many such entrepreneurs. Furniture production also saw the influence of Ford and the automobile industry. In the 1920's a large number of mechanical engineers in America banded together within the American Society of Mechanical Engineers in an effort to bring the woodworking industry into the twentieth century—into the century of mass production. Consequently, the ASME established in 1925 a Wood Industries Division, which served to focus the supposed great powers of mechanical engineering on all aspects of woodworking technology. In agriculture, Henry Ford himself argued that the problems of American agriculture could all be solved simply by adopting mass production techniques. Ford conducted experiments in this direction, but he was no more successful in agriculture than the mechanical engineers and housing fabricators were in bringing about mass production in their respective industries. One could argue, however, that today such an agricultural product as the hybrid tomato, bred to be picked, sorted, packaged, and transported by machinery, demonstrates that mass production methods have penetrated American agriculture. But in furniture and housing, there seems to be no equivalent to the hybrid tomato. . . .

. . . American furniture manufacturers continued to operate relatively small factories employing around 150 workers, annually turning out between 5,000 and 50,000 units. Beliefs that automotive production technology holds the key to abundance in all areas of consumption persist today. As recently as 1973, Richard Bender observed in his book on industrial building that "much of the problem of industrializing the building industry has grown out of the mistaken image of the automobile industry as a model." In many areas, the panacea of Fordism will continue to appeal to those who see in it solutions to difficult economic and social problems. The ethos of mass production, established largely by Ford, will die a hard death, if it ever disappears completely.

Yet the very timing of the rise of this ethos along with the appearance of the *Encyclopedia Britannica* article, "Mass Production," shows how full of paradox and irony history is. Although automotive America was rapidly growing in its consumption of everything under the sun and although Ford's achievements were known by an, mass production as Ford had made it and defined it was, for all intents and purposes, dead by 1926. Ford and his production experts had driven mass production into a deep cul-de-sac. American buyers had given up on the Ford Model T, and the Ford Motor Company watched its sales drop precipitously amid caustic criticism of its inability to accept and make changes. In mid-1927, Henry Ford himself finally gave up on the Model T after 15,000,000 of them had been produced. What followed in the changeover to the Model A was one of the most wrenching nightmares in American industrial history. Designing the new model, tooling up for its production, and achieving satisfactory production levels posed an array of unanticipated problems that led to a long delay in the Model A's introduction. In some respect, the Ford Motor Company never recovered from the effects of its first big changeover. Changes in consumers' tastes and gains in their disposable income made the Model T and the Model T idea obsolete. Automobile consumption in the late 1920s called for a new kind of mass production, a system which could accommodate frequent change and which was no longer wedded to the idea of maximum production at minimum cost. General Motors, not Ford, proved to be in tune with changes in American consumption with its explicit policy of "a car for every purpose and every purse," its unwritten policy of annual change, and its encouragement of "trading up" to a more expensive car. Ford learned painfully and at great cost that the times called for a new era, that of "flexible mass production."

The Great Depression dealt additional blows to Ford's version of mass production. With dramatic decreases in sales following the Great Crash, Ford and the entire industry began laying off workers. As a result, Detroit became known as the "beleaguered capital of mass production." Mass production had not prevented mass unemployment or, more properly, unemployment of the masses, but seemed rather to have exacerbated it. While overproduction had always posed problems for industrial economies, the high level of unemployment in the Great Depression made mass production an easy culprit for critics as they saw hundreds of thousands of men out of work in the Detroit area alone. Writing in the *New York Times* in 1931 Paul Mazur stressed that, "mass production has not proved itself to be an unmixed blessing; in the course of its onward march lie overproduction and the disastrous discontinuity of industry that comes as a consequence." Call it Fordism or mass production, it was nonetheless, "an alluring but false doctrine." Moreover, Mazur argued, "it is essential for business to realize that unquestioning devotion to mass production can [only] bring disaster.

Mazur's comments came in the wake of a previous *Times Magazine* article entitled, "Gandhi Dissects the Ford Idea." The article's author, Harold Callender, pitted Ford's doctrine of mass production against Mahatma Gandhi's notion that handicrafts, not mechanization, offered the solution to global problems of unemployment and hunger. The *Times Magazine* juxtaposed a photograph of an assembly line against one of a group of Indian hand spinners. Captions under the two photographs read as follows: "The Ford Formula for Happiness— A Mass-Production Line" and "The Gandi Formula for Happiness—A Group of Handicraft Spinners." While few would have agreed with the Gandhi, formula, Americans in the depths of the Depression certainly seem to have concluded that developments in mass production had not been matched by the development of "mass consumption." As Mazur put it, "the power of production . . . has been so great that its products have multiplied at geometric rates . . . at the same time the power of consumption—even under the influence of stimuli damned as unsocial and tending toward profligacy [e.g., advertising and built-in obsolescence (frequent style changes)]—has expanded only at a comparatively slow arithmetic rate."

While Americans may have had doubts about the doctrine of mass production, they by no means were willing to scrap it in favor of the Gandhi formula. Already their desire for style and novelty, coupled with increased purchasing power in the 1920s, had forced even Henry Ford to change his system of mass production. When pushed by the Depression, the greater part of Americans looked for solutions in the sphere of "mass consumption." The 1930s witnessed the publication of an extensive amount of literature on the "economics of consumption." As history would have it, the prophets of mass production were proven at least temporarily correct as the United States pulled itself out of the Depression by the mass consumption of war material and, after the war, by the golden age of American consumption in the 1950s and 1960s.

Section 2

Science, Technology and the World Wars

The Atomic Bombings Reconsidered

Reading 6

Barton J. Bernstein

■ The Questions America Should Ask

Fifty years ago, during a three-day period in August 1945, the United States dropped two atomic bombs on Japan, killing more than 115,000 people and possibly as many as 250,000, and injuring at least another 100,000. In the aftermath of the war, the bombings raised both ethical and historical questions about why and how they were used. Would they have been used on Germany? Why were cities targeted so that so many civilians would be killed? Were there likely alternative ways to end the war speedily and avoid the Allies' scheduled November 1, 1945, invasion of Kyushu?

Such questions often fail to recognize that, before Hiroshima and Nagasaki, the use of the A-bomb did not raise profound moral issues for policymakers. The weapon was conceived in a race with Germany, and it undoubtedly would have been used against Germany had the bomb been ready much sooner. During the war, the target shifted to Japan. And during World War II's brutal course, civilians in cities had already become targets. The grim Axis bombing record is well known. Masses of noncombatants were also intentionally killed in the later stages of the American air war against Germany; that tactic was developed further in 1945 with the firebombing of Japanese cities. Such mass bombing constituted a transformation of morality, repudiating President Franklin D. Roosevelt's prewar pleas that the warring nations avoid bombing cities to spare civilian lives. Thus, by 1945, American leaders were not seeking to avoid the use of the A-bomb on Japan. But the evidence from current archival research shows that by pursuing alternative tactics instead, they probably could still have obviated the dreaded invasion and ended the war by November.

■ Shifting from Germany to Japan

In 1941, urged by émigré and American scientists, President Roosevelt initiated the atomic bomb project—soon code-named the Manhattan Project—amid what was believed to be a desperate race with Hitler's Germany for the bomb. At the beginning, Roosevelt and his chief aides assumed that the A-bomb was a legitimate weapon that would be used first against Nazi Germany. They also decided that the bomb project should be kept secret from the Soviet Union, even after the Soviets became a wartime ally, because the bomb might well give the United States future leverage against the Soviets.

By mid-1944, the landscape of the war had changed. Roosevelt and his top advisers knew that the likely target would now be Japan, for the war with Germany would undoubtedly end well before the A-bomb was expected to be ready, around the spring of 1945. In a secret September 1944 memorandum at Hyde Park, Roosevelt and British Prime Minister Winston Churchill ratified the shift from Germany to Japan. Their phrasing suggested that, for the moment anyway, they might have had some slight doubts about actually using the bomb, for they agreed that "it might *perhaps*, after mature consideration, be used against the Japanese" (my emphasis).

Four days later, mulling over matters aloud with a visiting British diplomat and chief U.S. science adviser Vannevar Bush, Roosevelt briefly wondered whether the A-bomb should be dropped on Japan or whether it should be demonstrated in America, presumably with Japanese observers, and then used as a threat. His speculative notion seemed so unimportant and so contrary to the project's long-standing operating assumptions that Bush actually forgot about it when he prepared a memo of the meeting. He only recalled the president's remarks a day later and then added a brief paragraph to another memorandum.

Put in context alongside the dominant assumption that the bomb would be used against the enemy, the significance of F.D.R.'s occasional doubts is precisely that they were so occasional—expressed twice in almost four years. All of F.D.R.'s advisers who knew about the bomb always unquestioningly assumed that it would be used. Indeed, their memoranda frequently spoke of "after it is used" or "when it is used," and never "*if* it is used." By about mid-1944, most had comfortably concluded that the target would be Japan.

The bomb's assumed legitimacy as a war weapon was ratified bureaucratically in September 1944 when General Leslie Groves, the director of the Manhattan Project, had the air force create a special group—the 509th Composite Group with 1,750 men—to begin practicing to drop atomic bombs. So dominant was the assumption that the bomb would be used against Japan that only one high-ranking Washington official, Undersecretary of War Robert Patterson, even questioned this notion after V-E Day. He wondered whether the defeat of Germany on May 8, 1945, might alter the plans for dropping the bomb on Japan. It would not.

■ The Assumption of Use

The Manhattan Project, costing nearly $2 billion, had been kept secret from most cabinet members and nearly all of Congress. Secretary of War Henry L. Stimson, a trusted Republican, and General George C. Marshall, the equally respected army chief of staff, disclosed the project to only a few congressional leaders. They smuggled the necessary appropriations into the War Department budget without the knowledge—much less the scrutiny—of most congressmen, including most members of the key appropriations committees. A conception of the national interest agreed upon by a few men from the executive and legislative branches had revised the normal appropriations process.

In March 1944, when a Democratic senator heading a special investigating committee wanted to pry into this expensive project, Stimson peevishly described him in his diary as "a nuisance and pretty untrustworthy . . . He talks smoothly but acts meanly." That man was Senator Harry S Truman. Marshall persuaded him not to investigate the project, and thus Truman did not learn any more than that it involved a new weapon until he was suddenly thrust into the presidency on April 12, 1945.

In early 1945, James F. Byrnes, then F.D.R.'s "assistant president" for domestic affairs and a savvy Democratic politician, began to suspect that the Manhattan Project was a boondoggle. "If [it] proves a failure," he warned Roosevelt, "it will be subjected to relentless investigation and criticism." Byrnes' doubts were soon overcome by Stimson and Marshall. A secret War Department report, with some hyperbole, summarized the situation: "If the project succeeds, there won't be any investigation. If it doesn't, they won't investigate anything else."

Had Roosevelt lived, such lurking political pressures might have powerfully confirmed his intention to use the weapon on the enemy—an assumption he had already made. How else could he have justified spending roughly $2 billion, diverting scarce materials from other war enterprises that might have been even more useful, and bypassing Congress? In a nation still unprepared to trust scientists, the Manhattan Project could have seemed a gigantic waste if its value were not dramatically demonstrated by the use of the atomic bomb.

Truman, inheriting the project and trusting both Marshall and Stimson, would be even more vulnerable to such political pressures. And, like F.D.R., the new president easily assumed that the bomb should and would be used. Truman never questioned that assumption. Bureaucratic developments set in motion before he entered the White House reinforced his belief. And his aides, many inherited from the Roosevelt administration, shared the same faith.

■ Picking Targets

Groves, eager to retain control of the atomic project, received Marshall's permission in early spring 1945 to select targets for the new weapon. Groves and his associates had long recognized that they were considering a weapon of a new magnitude, possibly equivalent to the "normal bombs carried by [at least] 2,500 bombers." And they had come to assume that the A-bomb would be "detonated well above ground, relying primarily on blast effect to do material damage, [so that even with] minimum probable efficiency, there will be the maximum number of structures (dwellings and factories) damaged beyond repair."

On April 27, the Target Committee, composed of Groves, army air force men like General Lauris Norstad, and scientists including the great mathematician John Von Neumann, met for the first time to discuss how and where in Japan to drop the bomb. They did not want to risk wasting the precious weapon, and decided that it must be dropped visually and not by radar, despite the poor weather conditions in Japan during the summer, when the bomb would be ready.

Good targets were not plentiful. The air force, they knew, "was systematically bombing out the following cities with the prime purpose . . . of not leaving one stone lying on another: Tokyo, Yokohama, Nagoya, Osaka, Kyoto, Kobe, Yawata, and Nagasaki . . . The air force is operating primarily to laying [sic] waste all the main Japanese cities . . . Their existing procedure is to bomb the hell out of Tokyo."

By early 1945, World War II—especially in the Pacific—had become virtually total war. The firebombing of Dresden had helped set a precedent for the U.S. air force, supported by the American people, to intentionally kill mass numbers of Japanese citizens. The earlier moral insistence on noncombatant immunity crumbled during the savage war. In Tokyo, during March 9–10, a U.S. air attack killed about 80,000 Japanese civilians. American B-29S dropped napalm on the city's heavily populated areas to produce uncontrollable firestorms. It may even have been easier to conduct this new warfare outside Europe and against Japan because its people seemed like "yellow subhumans" to many rank-and-file American citizens and many of their leaders.

In this new moral context, with mass killings of an enemy's civilians even seeming desirable, the committee agreed to choose "large urban areas of not less than three miles in diameter existing in the larger populated areas" as A-bomb targets. The April 27 discussion focused on four cities: Hiroshima, which, as "the largest untouched target not on the 21st Bomber Command priority list," warranted serious consideration; Yawata, known for its steel industry; Yokohama; and Tokyo, "a possibility [though] now practically all bombed and burned out and . . . practically rubble with only the palace grounds left standing." They decided that other areas warranted more consideration: Tokyo Bay, Kawasaki, Yokohoma, Nagoya, Osaka, Kobe, Kyoto, Hiroshima, Kure, Yawata, Kokura, Shimonoseki, Yamaguchi, Kumamoto, Fukuoka, Nagasaki, and Sasebo.

The choice of targets would depend partly on how the bomb would do its deadly work—the balance of blast, heat, and radiation. At their second set of meetings, during May 11–12, physicist J. Robert Oppenheimer, director of the Los Alamos laboratory, stressed that the bomb material itself was lethal enough for perhaps a billion deadly doses and that the weapon would give off lethal radioactivity. The bomb, set to explode in the air, would deposit "a large fraction of either the initial active material or the radioactive products in the immediate vicinity of the target; but the radiation . . . will, of course, have an effect on exposed personnel in the target area." It was unclear, he acknowledged, what would happen to most of the radioactive material: it could stay for hours as a cloud above the place of detonation or, if the bomb exploded during rain or in high humidity and thus caused rain, "most of the active material will be brought down in the vicinity of the target area." Oppenheimer's report left unclear whether a substantial proportion or only a small fraction of the population might die from radiation. So far as the skimpy records reveal, no member of the Target Committee chose to dwell on this matter. They probably assumed that the bomb blast would claim most of its victims before the radiation could do its deadly work.

In considering targets, they discussed the possibility of bombing the emperor's palace in Tokyo and "agreed that we should not recommend it but that any action for this bombing should come from authorities on military policy." They decided to gather information on the effectiveness of using the bomb on the palace.

The Target Committee selected their four top targets: Kyoto, Hiroshima, Yokohama, and Kokura Arsenal, with the implication that Niigata, a city farther away from the air force 509th group's Tinian base, might be held in reserve as a fifth. Kyoto, the ancient former capital and shrine city, with a population of about a million, was the most attractive target to the committee. "From the psychological point of view," the committee minutes note, "there is the advantage that Kyoto is an intellectual center for Japan and [thus] the people there are more apt to appreciate the significance of such a weapon." The implication was that those in Kyoto who survived the A-bombing and saw the horror would be believed elsewhere in Japan.

Of central importance, the group stressed that the bomb should be used as a terror weapon—to produce "the greatest psychological effect against Japan" and to make the world, and the U.S.S.R. in particular, aware that America possessed this new power. The death and destruction would not only intimidate the surviving Japanese into pushing for surrender, but, as a bonus, cow other nations, notably the Soviet Union. In short, America could speed the ending of the war and by the same act help shape the postwar world.

By the committee's third meeting, two weeks later, on May 28, they had pinned down matters. They chose as their targets (in order) Kyoto, Hiroshima, and Niigata, and decided to aim for the center of each city. They agreed that aiming for industrial areas would be a mistake because such targets were small, spread on the cities' fringes, and quite dispersed. They also knew that bombing was imprecise enough that the bomb might easily miss its mark by a fifth of a mile, and they wanted to be sure that the weapon would show its power and not be wasted.

The committee understood that the three target cities would be removed from the air force's regular target list, reserving them for the A-bomb. But, the members were informed, "with the current and prospective rate of . . . bombings, it is expected to complete strategic bombing of Japan by 1 Jan 46 so availability of future [A-bomb] targets will be a problem." In short, Japan was being bombed out.

■ The Ratification of Terror Bombing

On May 28, 1945, physicist Arthur H. Compton, a Nobel laureate and member of a special scientific panel advising the high-level Interim Committee newly appointed to recommend policy about the bomb, raised profound moral and political questions about how the atomic bomb would be used. "It introduces the question of mass slaughter, really for the first time in history," he wrote. "It carries with it the question of possible radio-active poison over the area bombed. Essentially, the question of the use . . . of the new weapon carries much more serious implications than the introduction of poison gas."

Compton's concern received some independent support from General Marshall, who told Secretary Stimson on May 29 that the A-bomb should first be used not against civilians but against military installations—perhaps a naval base—and then possibly against large manufacturing areas after the civilians had received ample warnings to flee. Marshall feared "the opprobrium which might follow from an ill considered employment of such force." A graduate of Virginia Military Institute and a trained soldier, Marshall struggled to retain the older code of not *intentionally* killing civilians. The concerns of Compton the scientist and Marshall the general, their values so rooted in an earlier conception of war that sought to spare noncombatants, soon gave way to the sense of exigency, the desire to use the bomb on people, and the unwillingness or inability of anyone near the top in Washington to plead forcefully for maintaining this older morality.

On May 31, 1945, the Interim Committee, composed of Stimson, Bush, Harvard President James Conant, physicist and educator Karl T. Compton, Secretary of State designate James F. Byrnes, and a few other notables, discussed the A-bomb. Opening this meeting, Stimson, the aged secretary of war who had agonized over the recent shift toward mass bombing of civilians, described the atomic bomb as representing "a new relationship of man to the universe. This discovery might be compared to the discoveries of the Copernican theory and the laws of gravity, but far more important than these in its effects on the lives of men."

Meeting, as they were, some six weeks before the first nuclear test at Alamogordo, they were still unsure of the power of this new weapon. Oppenheimer told the group that it would have an explosive force of between 2,000 and 20,000 tons of TNT. Its visual effect would be tremendous. "It would be accompanied by a brilliant luminescence which would rise to a height of 10,000 to 20,000 feet," Oppenheimer reported. "The neutron effect [radiation] would be dangerous to life for a radius of at least two-thirds of a mile." He estimated that 20,000 Japanese would be killed.

According to the committee minutes, the group discussed "various types of targets and the effects to be produced." Stimson "expressed the conclusion, on which there was general agreement, that we could not give the Japanese any warning; that we could not concentrate on a civilian area; but that we should seek to make a profound psychological impression on as many of the inhabitants as possible. At the suggestion of Dr. Conant, the secretary agreed that the most desirable target would be a vital war plant employing a large number of workers and closely surrounded by workers' houses."

Directed by Stimson, the committee was actually endorsing terror bombing—but somewhat uneasily. They would not focus exclusively on a military target (the older morality), as Marshall had recently proposed, nor fully on civilians (the emerging morality). They managed to achieve their purpose—terror bombing—without bluntly acknowledging it to themselves. All knew that families—women, children, and, even in the daytime, during the bomb attack, some workers—dwelled in "workers' houses."

At the committee's morning or afternoon session, or at lunch, or possibly at all three times—different members later presented differing recollections—the notion of a noncombat demonstration of the A-bomb came up. The issue of how to use the bomb was not even on Stimson's agenda, nor was it part of the formal mandate of the Interim Committee, but he may have showed passing interest in the subject of a noncombat demonstration. They soon rejected it. It was deemed too risky for various reasons: the bomb might not work, the Japanese air force might interfere with the bomber, the A-bomb might not adequately impress the Japanese militarists, or the bomb might incinerate any Allied POWs whom the Japanese might place in the area.

The discussion on May 31 had focused substantially on *how* to use the bomb against Japan. At one point some of the members had considered trying several A-bomb strikes at the same time and presumably on the same city. Groves opposed this notion, partly on the grounds that "the effect would not be sufficiently distinct from our regular air force bombing program." Like the others, he was counting on the dramatic effect of a single bomb, delivered by a single plane, killing many thousands. It was not new for the air force to kill so many Japanese, but this method would be new. And the use of the new weapon would carry, as stressed by American proclamations in early August, the likelihood of more nuclear attacks on Japanese cities—a continuing "rain of ruin."

Two weeks after the Interim Committee meeting, on June 16, after émigré physicists James Franck and Leo Szilard and some colleagues from the Manhattan Project's Chicago laboratory raised moral and political questions about the surprise use of the bomb on Japan, a special four-member scientific advisory committee disposed of the matter of a noncombat demonstration. The group was composed of physicists Arthur Compton, J. Robert Oppenheimer, Enrico Fermi, and Ernest O. Lawrence. By one report, Lawrence was the last of the four to give up hope for a noncombat demonstration. Oppenheimer, who spoke on the issue in 1954 and was not then controverted by the other three men, recalled that the subject of a noncombat demonstration was not the most important matter dealt with during the group's busy weekend meeting and thus did not receive much attention. On June 16, the four scientists concluded: "We can propose no technical demonstration likely to bring an end to the war; we see no acceptable alternative to direct military use."

At that time, as some members of the scientific panel later grudgingly acknowledged, they knew little about the situation in Japan, the power of the militarists there, the timid efforts by the peace forces there to move toward a settlement, the date of the likely American invasion of Kyushu, and the power of the still untested A-bomb. "We didn't know beans about the military situation," Oppenheimer later remarked pungently.

But even different counsel by the scientific advisers probably could not have reversed the course of events. The bomb had been devised to be used, the project cost about $2 billion, and Truman and Byrnes, the president's key political aide, had no desire to avoid its use. Nor did Stimson. They even had additional reasons for wanting to use it: the bomb might *also* intimidate the Soviets and render them tractable in the postwar period.

Stimson emphasized this theme in a secret memorandum to Truman on April 25: "If the problem of the proper use of this weapon can be solved, we should then have the opportunity to bring the world into a pattern in which the peace of the world and our civilization can be saved." Concern about the bomb and its relationship to the Soviet Union dominated Stimson's thinking in the spring and summer of 1945. And Truman and Byrnes, perhaps partly under Stimson's tutelage, came to stress the same hopes for the bomb.

◼ The Agonies of Killing Civilians

During 1945, Stimson found himself presiding, with agony, over an air force that killed hundreds of thousands of Japanese civilians. Usually, he preferred not to face these ugly facts, but sought refuge in the notion that the air force was actually engaged in precision bombing and that somehow this precision bombing was going awry. Caught between an older morality that opposed the intentional killing of noncombatants and a newer one that stressed virtually total war, Stimson could neither fully face the facts nor fully escape them. He was not a hypocrite but a man trapped in ambivalence.

Stimson discussed the problem with Truman on June 6. Stimson stressed that he was worried about the air force's mass bombing, but that it was hard to restrict it. In his diary, Stimson recorded: "I told him I was anxious about this feature of the war for two reasons: first, because I did not want to have the United States get the reputation of outdoing Hitler in atrocities; and second, I was a little fearful that before we could get ready the air force might have Japan so thoroughly bombed out that the new weapon would not have a fair background to show its strength." According to Stimson, Truman "laughed and said he understood."

Unable to reestablish the old morality and wanting the benefits for America of the new, Stimson proved decisive—even obdurate—on a comparatively small matter: removing Kyoto from Groves' target list of cities. It was not that Stimson was trying to save Kyoto's citizens; rather, he was seeking to save its relics, lest the Japanese become embittered and later side with the Soviets. As Stimson explained in his diary entry of July 24: "The bitterness which would be caused by such a wanton act might make it impossible during the long post-war period to reconcile the Japanese to us in that area rather than to the Russians. It might thus . . . be the means of preventing what our policy demanded, namely, a sympathetic Japan to the United States in case there should be any aggression by Russia in Manchuria."

Truman, backing Stimson on this matter, insisted privately that the A-bombs would be used only on military targets. Apparently the president wished not to recognize the inevitable—that a weapon of such great power would necessarily kill many civilians. At Potsdam on July 25, Truman received glowing reports of the vast destruction achieved by the Alamogordo blast and lavishly recorded the details in his diary: a crater of 1,200 feet in diameter, a steel tower destroyed a half mile away, men knocked over six miles away. "We have discovered," he wrote in his diary, "the most terrible bomb in the history of the world. It may be the fire destruction prophesied." But when he approved the final list of A-bomb targets, with Nagasaki and Kokura substituted for Kyoto, he could write in his diary, "I have told Sec. of War . . . Stimson to use it so that military objectives and soldiers and sailors are the target and not women and children. Even if the Japs are savages, ruthless, merciless, and fanatic . . . [t]he target will be a purely military one." Truman may have been engaging in self-deception to make the mass deaths of civilians acceptable.

Neither Hiroshima nor Nagasaki was a "purely military" target, but the official press releases, cast well before the atomic bombings, glided over this matter. Hiroshima, for example, was described simply as "an important Japanese army base." The press releases were drafted by men who knew that those cities had been chosen partly to dramatize the killing of noncombatants.

On August 10, the day after the Nagasaki bombing, when Truman realized the magnitude of the mass killing and the Japanese offered a conditional surrender requiring continuation of the emperor, the president told his cabinet that he did not want to kill any more women and children. Rejecting demands to drop more atomic bombs on Japan, he hoped not to use them again. After two atomic bombings, the horror of mass death had forcefully hit the president, and he was willing to return partway to the older morality—civilians might be protected from A-bombs. But he continued to sanction the heavy conventional bombing of Japan's cities, with the deadly toll that napalm, incendiaries, and other bombs produced. Between August 10 and August 14—the war's last day, on which about 1,000 American planes bombed Japanese cities, some delivering their deadly cargo after Japan announced its surrender—the United States probably killed more than 15,000 Japanese.

■ The Roads Not Taken

Before August 10, Truman and his associates had not sought to avoid the use of the atomic bomb. As a result, they had easily dismissed the possibility of a noncombat demonstration. Indeed, the post-Hiroshima pleas of Japan's military leaders for a final glorious battle suggest that such a demonstration probably would not have produced a speedy surrender. And American leaders also did not pursue other alternatives: modifying their unconditional surrender demand by guaranteeing the maintenance of the emperor, awaiting the Soviet entry into the war, or simply pursuing heavy conventional bombing of the cities amid the strangling naval blockade.

Truman and Byrnes did not believe that a modification of the unconditional surrender formula would produce a speedy surrender. They thought that guaranteeing to maintain the emperor would prompt an angry backlash from Americans who regarded Hirohito as a war criminal, and feared that this concession might embolden the Japanese militarists to expect more concessions and thus prolong the war. As a result, the president and his secretary of state easily rejected Stimson's pleas for a guarantee of the emperor.

Similarly, most American leaders did not believe that the Soviet entry into the Pacific war would make a decisive difference and greatly speed Japan's surrender. Generally, they believed that the U.S.S.R.'s entry would help end the war—ideally, before the massive invasion of Kyushu. They anticipated Moscow's intervention in mid-August, but the Soviets moved up their schedule to August 8, probably because of the Hiroshima bombing, and the Soviet entry did play an important role in producing Japan's surrender on August 14. Soviet entry without the A-bomb *might* have produced Japan's surrender before November.

The American aim was to avoid, if possible, the November 1 invasion, which would involve about 767,000 troops, at a possible cost of 31,000 casualties in the first 30 days and a total estimated American death toll of about 25,000. And American leaders certainly wanted to avoid the second part of the invasion plan, an assault on the Tokyo plain, scheduled for around March 1, 1946, with an estimated 15,000–21,000 more Americans dead. In the spring and summer of 1945, no American leader believed—as some later falsely claimed—that they planned to use the A-bomb to save half a million Americans. But, given the patriotic calculus of the time, there was no hesitation about using A-bombs to kill many Japanese in order to save the 25,000–46,000 Americans who might otherwise have died in the invasions. Put bluntly, Japanese life—including civilian life—was cheap, and some American leaders, like many rank-and-file citizens, may well have savored the prospect of punishing the Japanese with the A-bomb.

Truman, Byrnes, and the other leaders did not have to be reminded of the danger of a political backlash in America if they did not use the bomb and the invasions became necessary. Even if they had wished to avoid its use—and they did not—the fear of later public outrage spurred by the weeping parents and loved ones of dead American boys might well have forced American leaders to drop the A-bomb on Japan.

No one in official Washington expected that one or two atomic bombs would end the war quickly. They expected to use at least a third, and probably more. And until the day after Nagasaki, there had never been in their thinking a choice between atomic bombs and conventional bombs, but a selection of both—using mass bombing to compel surrender. Atomic bombs and conventional bombs were viewed as supplements to, not substitutes for, one another. Heavy conventional bombing of Japan's cities would probably have killed hundreds of thousands in the next few months, and might have produced the desired surrender before November 1.

Taken together, some of these alternatives—promising to retain the Japanese monarchy, awaiting the Soviets' entry, and even more conventional bombing—very probably could have ended the war before the dreaded invasion. Still, the evidence—to borrow a phrase from F.D.R.—is somewhat "iffy," and no one who looks at the intransigence of the Japanese militarists should have full confidence in those other strategies. But we may well regret that these alternatives were not pursued and that there was not an effort to avoid the use of the first A-bomb—and certainly the second.

Whatever one thinks about the necessity of the first A-bomb, the second—dropped on Nagasaki on August 9—was almost certainly unnecessary. It was used because the original order directed the air force to drop bombs "as made ready" and, even after the Hiroshima bombing, no one in Washington anticipated an imminent Japanese surrender. Evidence now available about developments in the Japanese government—most notably the emperor's then-secret decision shortly before the Nagasaki bombing to seek peace—makes it clear that the second bomb could undoubtedly have been avoided. At least 35,000 Japanese and possibly almost twice that number, as well as several thousand Koreans, died unnecessarily in Nagasaki.

Administration leaders did not seek to avoid the use of the A-bomb. They even believed that its military use might produce a powerful bonus: the intimidation of the Soviets, rendering them, as Byrnes said, "more manageable," especially in Eastern Europe. Although that was not the dominant purpose for using the weapon, it certainly was a strong confirming one. Had Truman and his associates, like the dissenting scientists at Chicago, foreseen that the A-bombing of Japan would make the Soviets intransigent rather than tractable, perhaps American leaders would have questioned their decision. But precisely because American leaders expected that the bombings would also compel the Soviet Union to loosen its policy in Eastern Europe, there was no incentive to question their intention to use the atomic bomb. Even if they had, the decision would probably have been the same. In a powerful sense, the atomic bombings represented the implementation of an assumption—one that Truman comfortably inherited from Roosevelt. Hiroshima was an easy decision for Truman.

▪ The Redefinition of Morality

Only years later, as government archives opened, wartime hatreds faded, and sensibilities changed, would Americans begin seriously to question whether the atomic bombings were necessary, desirable, and moral. Building on the postwar memoirs of Admiral William Leahy and General Dwight D. Eisenhower, among others, doubts began to emerge about the use of the atomic bombs against Japan. As the years passed, Americans learned that the bombs, according to high-level American military estimates in June and July 1945, probably could not have saved a half million American lives in the invasions, as Truman sometimes contended after Nagasaki, but would have saved fewer than 50,000. Americans also came slowly to recognize the barbarity of World War II, especially the mass killings by bombing civilians. It was that redefinition of morality that made Hiroshima and Nagasaki possible and ushered in the atomic age in a frightening way.

That redefinition of morality was a product of World War II, which included such barbarities as Germany's systematic murder of six million Jews and Japan's rape of Nanking. While the worst atrocities were perpetrated by the Axis, all the major nation-states sliced away at the moral code—often to the applause of their leaders and citizens alike. By 1945 there were few moral restraints left in what had become virtually a total war. Even F.D.R.'s prewar concern for sparing enemy civilians had fallen by the wayside. In that new moral climate, any nation that had the A-bomb would probably have used it against enemy peoples. British leaders as well as Joseph Stalin endorsed the act. Germany's and Japan's leaders surely would have used it against cities. America was not morally unique—just technologically exceptional. Only it had the bomb, and so only it used it.

To understand this historical context does not require that American citizens or others should approve of it. But it does require that they recognize that pre- and post-Hiroshima dissent was rare in 1945. Indeed, few then asked why the United States used the atomic bomb on Japan. But had the bomb not been used, many more, including numerous outraged American citizens, would have bitterly asked that question of the Truman administration.

In 1945, most Americans shared the feelings that Truman privately expressed a few days after the Hiroshima and Nagasaki bombings when he justified the weapons' use in a letter to the Federal Council of Churches of Christ. "I was greatly disturbed over the unwarranted attack by the Japanese on Pearl Harbor and their murder of our prisoners of war," the president wrote. "The only language they seem to understand is the one we have been using to bombard them. When you have to deal with a beast you have to treat him as a beast."

Nazi Science and Nazi Medical Ethics: Some Myths and Misconceptions

Reading 7

Robert N. Proctor

We often hear that the Nazis destroyed science and abandoned ethics. That was the view of Telford Taylor in his opening statement at the Nuremberg "Doctor's Trial" of 1946–1947, where he stated that the Nazi doctors had turned Germany "into an infernal combination of a lunatic asylum and a charnel house" where "neither science, nor industry, nor the arts could flourish in such a foul medium" [1]. Similar views were expressed by Franz Neumann, author of the 1942 treatise *Behemoth*, the first major analysis of how the Nazis came to power [2]. Neumann predicted "a most profound conflict" between the "magic character" of Nazi propaganda and the "rational" processes of German industry, a conflict the emigré political theorist believed would culminate in an uprising on the part of engineers to combat the irrationalist regime. Such an uprising, needless to say, never materialized.

It would be comforting to believe, of course, that good science tends to travel with good ethics, but the sad truth seems to be that cruelty can coexist fairly easily with "good science." There is a convenience of sorts in the myth: it makes it easy to argue that "Nazi science" was not really science at all, and therefore there is no ethical dilemma. One needn't talk about the ethics of Nazi medicine, since there was no legitimate medicine to speak of. Nazi science in one swift blow is reduced to an oxymoron, a medical non-problem.[1]

Myth-Making and Exculpation

Different groups have participated in this process of myth-making; let me mention four.

First, for *Germans who stayed in Germany*, the myth of suppressed science served as a way for post-war scientists to distance themselves from their Nazi past, to block historical investigators from dredging up potentially embarrassing collaborations. If there was no such thing as Nazi science, what is there to investigate? As an epidemiologist at Germany's National Cancer Center in Heidelberg once confessed: "1945 was a scientific *Stunde Null*, we don't look at what came before." This allowed scientists to argue that the very act of scientific research under Nazi rule was a form of resistance. It was often advertised as such, after the war.

Second, for *Jewish scholars forced from Germany*, there was also the unwillingness to believe that the system that had treated them so shoddily had continued to produce good science. This view was reinforced by the fact that fields in which Jews had been prominent (biochemistry and quantum mechanics, for example, but also medical specialties like dermatology) tended to be the fields most heavily gutted by Nazi policies. Fields with lesser Jewish representation—like veterinary medicine or surgery—generally speaking suffered less and therefore have drawn less critical historical attention because there were fewer emigrés in these fields [4].

Third, for *American military authorities*, the myth of flawed science also served to disguise the fact that even as U.S. officials were denouncing Nazi science, they were also busily trying to recruit Nazi talent for use in U.S. military projects. At least 1,600 German scientists came to the United States under the rubric of "Operation Paperclip"—including not just SS officers like Werner von Braun, but also a number of medical professionals, some of whom had been implicated in abusive human experimentation [5].

---

[1]Analyzing the notorious Dachau hypothermia experiments, Robert Berger claimed that "the Nazi data" were not scientific, any ethical debate over whether to use Nazi science was misconceived from the outset; see [3].

"Nazi Science and Nazi Medical Ethics: Some Myths and Misconceptions" by Robert N. Proctor. Perspectives in Biology & Medicine *43:3 (2000), 335–346. © 2000 The Johns Hopkins University Press. Reprinted with permission of The Johns Hopkins University Press.*

Finally, this myth, as I have identified it, served to reassure *the American public* that abuses like those of the Nazi era could never occur in a liberal democracy. Nazi science was pseudo-science, science out-of-control; American science was genuine science, secure within democratic institutions, obedient to the rule of law. Post-war ethical codes of conduct could even be dismissed as unnecessary—after all, weren't they designed to prevent abuses that could only occur in a totalitarian society? That, apparently, was the thinking of some of those involved in the radiation experiments investigated by the President's Advisory Committee on Human Radiation Experiments: for example, the Vanderbilt nutritionist William J. Darby, who from 1945 through 1947 supervised the feeding of radioactive iron to 829 pregnant women without their consent, when asked whether he recognized the legitimacy of the Nuremberg Code, responded that he did not, because the code applied only to "experiments of a medical nature committed by the Germans . . . a different setting entirely" [6].

Triumphs in Science

Germany was not a technical or medical backwater in the 1930s and 1940s. Anyone who has ever examined a V-2 engine will have little doubts about this, but there are many other examples. Nazi-era scientists and engineers were pioneers of television, jet-propelled aircraft, guided missiles, electronic computers, the electron microscope and ultracentrifuge, atomic fission, new data processing technologies, new pesticides, and the world's first industrial murder factories (including the use of gas chambers disguised as showers)—all of which were either first developed in Nazi Germany or reached their high point at that time. We should recall that the first magnetic tape recording was of a speech by Hitler, that the V-2 emerged from a plan for intercontinental ballistic missiles designed to be able to reach New York City, and that the nerve gases sarin and tabun were Nazi inventions. I have recently published a book arguing that German cancer research at this time was the most advanced in the world: Nazi-era health reformers built on this research base, introducing smoke-free public spaces, bans on carcinogenic food dyes, and new means of controlling dust exposure on factory floors. The period saw extensive work in the area of occupational carcinogenesis, and in 1943, Germany became the first nation to recognize lung cancer and mesothelioma as compensable occupational illnesses caused by asbestos inhalation [see 7, 8].

The story of science under German fascism must therefore be more than a narrative of suppression and survival; it must also tell how Nazi ideology *promoted* certain areas of inquiry, how projects and policies came and went with the movement of political forces. The frightening truth is that the Nazis supported many kinds of science, left politics (as we often think about it) out of most, and transformed but did not abandon ethics. There is an ethics of Nazi medical practice, sometimes explicit, sometimes not; sometimes cruel, sometimes not. This is important to understand if we are not to demonize the Nazi phenomenon as something absolutely alien and otherworldly.

Appreciating Nazi support for science and medicine can help us understand the appeal of Nazism within German intellectual culture. I also want to argue, though, that what went wrong in the Nazi period is not best understood as a subordination of the good of the individual to the good of the whole. Rather, one has to understand who was included within "the whole" and who was banished. It was not, after all, "the individual" in the abstract who suffered but particular kinds of individuals. Public health protections were extended to the "healthy" majority, while so-called "enemies of the people" were first excluded and then exterminated. Medicine was complicit in both ends of this moral scale: in public health reforms that brought the majority of Germans cleaner air and water, and in "health reforms" that involved sterilization and eventually wholesale murder.

The Example of Tobacco Research

If you ask most people when the first good evidence arose that tobacco was a major cause of lung cancer, they will point to a series of epidemiological studies by English and Americans in the early 1950s. If you ask when a medical consensus on this question first arose, they will most likely point to the 1964 Surgeon General's report, which took a strong stand on this question, or a similar report by Britain's Royal College of Physicians two years earlier [9, 10].

I have become convinced, however, that there is an earlier and overlooked consensus during the Nazi period. The Nazis had a powerful anti-tobacco movement, arguably the most powerful in the world at that time. Tobacco was opposed by racial hygienists who feared the corruption of the German germ plasm, by industrial hygienists who feared a reduction of work capacity, and by nurses and midwives who feared harms for the "maternal organism." Tobacco was said to be "a corrupting force in a rotting civilization that has become lazy." The Nazi-era anti-tobacco rhetoric drew from an earlier generation's eugenic rhetoric, combining this with an ethic of bodily purity and performance at work [11]. Tobacco use was attacked as "epidemic," as a "plague," as "dry drunkenness," and as "lung masturbation"; tobacco and alcohol abuse were "diseases of civilization" and "relics of a liberal lifestyle" [12].

Anti-tobacco research also flourished in the Nazi period. Animal experimental work demonstrated that the tar extracted from cigarette smoke could cause cancer, and physical chemists distilled tobacco tars to identify the carcinogenic components. The editor of Germany's *Monatsschrift für Krebsbekämpfung* organized animal experiments to test whether smoking causes lung cancer, putting rats in a "gas chamber" with cigarette smoke pumped in from the top—until the animals suffocated. The description of the dying rats, gasping and falling over one another, is chilling, given subsequent events [13].

Germans also pioneered what we now call experimental tobacco epidemiology, the two most striking papers being a 1939 article by Franz H. Müller of Cologne's City Hospital, and a 1943 paper by Eberhard Schairer and Erich Schöniger at Jena, presenting the most convincing demonstrations up to that time that cigarettes were a major cause of lung cancer [14, 15]. In his paper, Müller analyzed the smoking habits of 83 male lung cancer patients and compared these with the habits of age-standardized "controls" not suffering from lung cancer. His findings were clear-cut and striking: the lung cancer patients were much more likely to be heavy smokers and much less likely to be non-smokers. Sixteen percent of the healthy group were non-smokers, compared with only 3.5 percent for the lung cancer group. The 86 lung cancer patients smoked a total of 2,900 grams of tobacco per day, while the 86 healthy men smoked only 1,250 grams. Müller concluded that tobacco was not just "an important cause" of lung cancer, but also that "*the extraordinary rise in tobacco use*" was "*the single most important cause of the rising incidence of lung cancer*" in recent decades [14, emphasis in original].

Müller's work was taken one step further by Schairer and Schöniger, two physicians working at Jena's Institute for Tobacco Hazards Research. The authors were aware that German lung cancer rates were on the rise, and that many of the non-tobacco explanations of the rise were flawed (the automotive exhaust theory, for example, failed to explain the fact that rural rates were also rising). The authors drew attention to the fact that a heavy smoker could inhale as much as four kilograms of tar over a lifetime, a frightening figure given Angel H. Roffo's demonstration that animals painted with tobacco tars develop high rates of cancer [15].

Following closely the method pioneered by Müller, Schairer and Schöniger sent questionnaires to the relatives of 195 lung cancer victims, inquiring into the smoking habits of the deceased. An additional 555 questionnaires were sent to the families of patients who had died from other kinds of cancer—the presumption being that smokers would be more likely to develop certain kinds of cancer rather than others. Questionnaires were also sent to 700 male residents of Jena to determine the smoking habits among a population apparently free of cancer. The results again were clear: among the 109 lung cancer cases for which useable data were obtained, only three were non-smokers, a far lower proportion than among the population as a whole (about 3 percent, versus 16 percent for the non-cancer controls). The smokers were not necessarily "cancer prone," because when other kinds of cancer were looked at—stomach, for example—smokers were found to be no more likely to develop cancer than non-smokers. Their conclusion: smoking was very likely a cause of lung cancer, but much less likely a cause of other kinds of cancer. The results were of the "highest" statistical significance; a 1994 reevaluation of the study showed that the probability that the results could have come about by chance was less than one in ten million [16].

■ Questions of Interpretation

How do we interpret such studies? How do we explain the fact that Nazi Germany was home to the world's foremost tobacco-cancer epidemiology, the world's strongest cancer prevention policy, or the world's first recognition that asbestos could cause lung cancer? Do we say that "pockets of innovation" existed in Nazi Germany, resistant to ideological influence?[2] What if we find, on closer inspection, that Germany's anti-tobacco research flourished not just *despite* the Nazis, but in large part *because of* the Nazis? Is it kosher then to cite such studies?

---

[2]On the "pockets of innovation" thesis, see [17, 18].

I ask this last question, partly because the tobacco studies mentioned above have, in fact, been occasionally cited, though rarely with any mention of the social context within which they arose. (A notable exception is [19].) There is never any mention, for example, of the fact that the founding director of Schöniger and Schairer's Institute was Karl Astel, president of the University of Jena and a vicious racial hygienist and SS officer. There is never a mention of the fact that the grant application for the Institute was written by Gauleiter Fritz Sauckel, chief organizer of Germany's system of forced labor, a man hanged after the war for crimes against humanity. (Most leaders of Germany's anti-tobacco movement were silenced in one way or another after Germany's defeat. Müller disappeared in the war, Hans Reiter lost his job, Astel and Leonardo Conti committed suicide, etc.) No mention is ever made of the fact that funding for Astel's institute came from a RM 100,000 personal gift from the Führer, himself an ardent anti-smoking activist. It is clear to anyone who follows the money trail that Schairer and Schöniger's study would not have been undertaken had it not been for Hitler's interest, and the interest of several of his underlings. I might also point out that Hitler at one point even attributed the rise of German fascism to his quitting smoking: the young and struggling artist smoked a couple of packs a day until 1919, at which time he threw his cigarettes into the Danube and never reached for them again. In a 1942 conversation, the Führer said that Germany owed its liberation (i.e., the triumph of Nazism) to his quitting smoking—though I guess one could say, contra Hitler and with a hint of Freud, that sometimes giving up smoking is just giving up smoking.

What do we make of the fact that Nazi ideology in this case (and there are others) appears not just not to have hindered research, but actually to have promoted it?

In drawing attention to such studies, my intention is not to argue that there was "something good" that came from the Nazi era. I have no desire to rescue the honor of this era, or to "balance the historical record" for balance's sake. You cannot balance the heavy weight of genocide against a few flashy epidemiological studies or a glitzy V-2 engine. That is not the point—and I should add that I have little sympathy for those who argue that republishing brutal Nazi experiments may mean, as some have argued, that the victims of such experiments "may not have died in vain." They did die in vain, and a better designed life preserver or Saturn V rocket is no compensation.

My point is, rather, that it does us little good to caricature Nazi medicine as irrational or anti-science *in the abstract*. What we have to look at more carefully is the relationship between science and ideology at this time. I do not believe, for example, that the papers on tobacco epidemiology I have cited were uninfluenced by Nazi ideology. The anti-tobacco program was motivated by Nazi ideals of bodily purity and racial hygiene: there is a kind of "homeopathic paranoia" pervading Nazi ideology that led many to believe that tiny, corrosive poisons were infiltrating the German *Volkskörper*, sapping its strength, causing harm. Appreciating this helps us understand how Nazi medical ideologues could argue that lead, mercury, asbestos, Jews, and tobacco tars were all a menace to the German body. I don't believe that doctors were "doubling" when they sought to cleanse the German population of those elements; nor do I believe that we can identify a sharp boundary between the science and the politics in this process. The two are painfully intertwined.

Why were German doctors such avid fans of fascism? I don't think it was the tirades of Julius Streicher in *Der Stürmer* that caught their eye, but rather the promises of Nazi leaders to solve Germany's problems medically, surgically. The Nazi state was supposed to be a hygienic state; Nazism was supposed to be "applied biology." (Fritz Lenz coined this widely cited phrase in his popular human genetics textbook of 1931 [20].) Hitler was celebrated as the "Robert Koch of politics" and the "great doctor" of German society. The seductive power of National Socialism lay in its promise to cleanse German society of its corrosive elements—not just communism and Jews, but pollutants in the air and water, along with TB, homosexuality, and the "burdensome" mentally ill—maladies which could all be traced, in the Nazi view of the world, to the "false humanitarianism" of previous political regimes. The doctors in this process were not victims but coconspirators, seduced by the powers being offered to them and the promises of an orderly, hygienic state.

■ Nazi Medical Ethics

We often hear that the Nazis abandoned ethics. A recent Israeli film on the euthanasia operation says that Nazi doctors' overzealous scientific curiosity led them "to abandon all moral sense in the pursuit of medical knowledge" [21]. The image is of the unfettered quest for knowledge, a kind of scientific zealotry reminiscent of the Faustian bargain, of science practiced "without limits" or of an overly "aggressive search for the truth."

The problem with this view is that there were in fact ethical standards at this time. Medical students took courses on medical ethics, and medical textbooks from the time treated medical ethics.[3] There are discussions in German journals of the obligations of physicians to society, to the state, and sometimes even to the individual. Nazi medical philosophers were critical of the ideal of "neutral" or value-free science, which was often equated with apathetic ivory-tower liberal-Jewish "science for its own sake." Science was supposed to serve the German Volk, the healthy and productive white races of Europe. We have to distinguish between no ethics and a lot of bad ethics, between chaos and evil.

Surprisingly, there never has been a systematic study of medical ethics under the Nazis. We know a lot about the postwar rationalizations of physicians facing death at the hands of Allied prosecutors [24, pp. 133–34], but we need to know more about the implicit and explicit norms of the era; we need to know more about bedside manners and student-teacher relations, and what kinds of things were discussed when medical malpractice suits went to court. Such questions are especially intriguing, given that Germany prior to the Nazi era had taken steps to protect patients' rights in the experimental context. The 1900 code promulgated by the Prussian Ministry of Religion, Education, and Medical Affairs, for example, was the world's first official regulation of human experimentation, barring non-therapeutic interventions without voluntary consent, along with experiments on minors and others judged vulnerable or incompetent. Experiments had to be authorized by the director of the institution involved, and records had to be kept in writing [24, p. 127].

A 1931 code issued by the Reich Health Office strengthened sanctions against inappropriate human experiments. The edict was issued in response to a growing public clamor about experimental abuses—most notably the deaths of 75 children in Lübeck in the course of testing an experimental TB vaccine. The 1931 guidelines banned experiments on the dying and on anyone under the age of 18 if it posed a risk. The document has been called the world's most comprehensive code governing human experimentation; in certain respects it is stricter even than the subsequent Nuremberg Code or Helsinki Accord [24, pp. 121–44].

Medical ethical discourse continued throughout the Nazi period. On 24 November 1933, for example, a law for the protection of cruelty to animals was passed, barring experimentation causing pain or injury to animals. The law specifically disallowed experiments involving exposure to cold, heat, or infection. Hermann Göring threatened to toss vivisectionists into a concentration camp, though it is unclear whether this ever took place. Jewish doctors were commonly accused of excessive experimental zeal: Erwin Liek, a kind of spiritual godfather for many Nazi physicians, linked abuses in this sphere with the "Jewish abstract scientific spirit" [25, 26].

There was also a great deal of discussion of physicians' duties and responsibilities—especially their obligations to serve ("unflinchingly") the Volk and the Führer. Doctors were instructed to counsel their patients against tobacco use [27], and were enlisted in the task of safeguarding public and genetic health. The job of maintaining labor productivity fell to physicians, who also were asked to devote increasing attention to neonatal care. There are discussions of who should be allowed to practice medicine—whether natural healers were to be barred from treating cancer patients, for example (they were) [28]—and there are discussions of the limits of medical confidentiality and of medical disclosure. A 1943 article in a leading cancer journal cited the "demands of medical ethics" to inform patients of the severity of their diseases, and in at least one case a physician was prosecuted for failing to inform a woman she had cancer [29].

We also know that doctors did not stand idly by when they thought things were going wrong. There was a culture of complaint in the Nazi period: doctors complained that sterilizations were going too fast or too slow, that too many people were being X-rayed or too few, and so forth. Physicians complained about the use of fancy-sounding obfuscatory Latin to characterize diseases (doctors were supposed to use plain German, so patients could understand the nature of their illness), and thousands of women were said to be needlessly dying from lack of access to preventive care (e.g., regular, cost-free gynecologic exams, proposed by many Nazi doctors) [30, 31].

The Nazi medical profession, in other words, was not without its ethics. Nazi medical ethics was typically sexist and racialist, but there was also stress on cleanliness, punctuality, orderliness, and obedience to legal authority, especially the supreme authority of Adolf Hitler, the beloved Führer. Nazi medical ethics tended

---

[3]An example of Nazi medical ethical discussion is [22]. The most widely cited medical ethics text at this time was [23].

to reduce morality to efficiency, economics, and aesthetics, relegating to the trash heap everything that was seen as ugly and a burden. Preventive medicine was emphasized, as was Nordic (and especially Germanic) supremacy, cost-efficiency, the natural lifestyle, and the superiority of the productive worker over persons deemed inferior or burdensome.

Ethical norms are implicit even in the most horrific experimental practices of the camps: how else does one explain the fact that "healthy" German citizens were never experimented on? Those subjected to human experimental violence were invariably people judged less than fully human in the Nazi scale of values. Jews and Gypsies were considered diseased races, tumors in the German body politic; Russian POWs were vermin deserving enslavement or extermination; the unproductive handicapped were encumbrances on the German Volk, warranting "euthanasia."

It is important not to make a caricature of Nazi ideology. The doctors who exploited prisoners at Buchenwald or Dachau were not morally blind or devoid of the power of moral reflection. Upholding such a view would almost render the guilty parties not responsible for their actions: you cannot hold guilty parties accountable if they are lunatics devoid of morals. Some of the worst were thinking men, acting consistently within a frame of callous and often criminal logic.

■ Research Integrity versus Research Ethics

The U.S. Office of Research Integrity (ORI) defines "scientific misconduct" to include "fabrication, falsification . . . or other practices that seriously deviate from those that are commonly accepted within the scientific community for proposing, conducting, or reporting research." The NSF similarly defines misconduct as "fabrication, falsification, plagiarism, or other serious deviation from accepted practices in proposing, carrying out, or reporting results from activities funded by the NSF" [32]. What is interesting about these definitions is that they avoid the difficult issue of how one deals with large-scale, deeply ingrained prejudices that may be widely accepted in both the scientific community and the society as a whole. Narrow definitions of scientific misconduct say nothing about whether a given research practice may be abusive, or racist, or sexist—nothing about its larger context or implications.

Some of the worst Nazi research cannot even be considered misconduct according to such definitions. Most Nazi doctors did not lie, or cheat, or misrepresent their credentials. They did not falsify or fabricate data to an unusual degree, and there is little evidence of plagiarism. The evil must be sought elsewhere.

The primary failing of Nazi medicine, I would argue, was the failure of physicians to challenge the rotten, substantive core of Nazi values. Too many physicians were willing to go with the political flow; too many were unwilling to resist, to "deviate" from "commonly accepted" practices. The ORI definition of "misconduct" misses this larger point entirely. Ian Kershaw reminds us that "the road to Auschwitz was paved with indifference"; I would only add that conformist urges blocked the many side paths leading in other directions. Doctors should have acted up, broken with convention, defended their patients against the long arm of the law.

■ The Black in White, the White in Black

My point is not to rescue the honor of medicine in its darkest hour, but rather to stress its subtlety and complexity. The history of medicine in this period is a history of both forcible sterilization and herbal remedies; we cannot forget the crimes of a Karl Brandt or a Hermann Voss [33], but we also should not forget that the SS built the world's largest botanical medical garden in Dachau, or that German nutritionists mandated the production of whole-grain bread. Fascist physicians willfully murdered their handicapped patients, but organic farming and species protection were also going concerns. The question is not one of balance, but of the proper understanding of origins, context, continuities, and contradictions. It is part of the horror of this period, that such an "advanced" technological society could fall so far into butchery and barbarism.

I do not believe there is an inherently totalitarian tendency in modern science, but I do think it is important to recognize that, just as the routine practice of science is not incompatible with the routine exercise of cruelty, so the dictatorial and murderous aspirations of fascism were not necessarily at odds with the promotion of cutting edge science and progressive public health—at least for certain elements of the population. The exclusive focus on the heinous aspects of Nazi medical practice makes it easy for us to relegate the events of this era to the monstrous or otherworldly, but there is more to the story than "medicine gone mad." The Nazi campaign against carcinogenic food dyes, the world-class asbestos and tobacco epidemiology, and much else as well, are all in some sense as fascist as the yellow stars and the death camps. There is sometimes white in black, and black in white; appreciating some of these subtler speckles and shadings may open our eyes to new kinds of continuities binding the past to the present. It may also help us better see how fascism triumphed in the first place.[4]

References

1. Taylor, T. Opening statement of the prosecution, December 9, 1946. In *The Nazi Doctors and the Nuremberg Code*, edited by G. J. Annas and M. A. Grodin. New York: Oxford Univ. Press, 1992. 69.
2. Neumann, F. *Behemoth: The Structure and Practice of National Socialism*. New York: Oxford Univ. Press, 1942. 471–72.
3. Berger, R. Nazi science: The Dachau hypothermia experiments. *New Engl. J. M.* 322:1435–40, 1990.
4. Fischer, W., et al., eds. *Exodus von Wissenschaften aus Berlin: Fragestellung, Ergebnisse, Desiderate*. Berlin: de Gruyter, 1994.
5. Hunt, L. *Secret Agenda: The United States Government, Nazi Scientists, and Project Paperclip, 1945–1990*. New York: St. Martin's Press, 1991.
6. Sea, G. Time distance, shielding: Human radiation experiments and the laying of blame. Paper presented to the Conference on Humans in Experiments, Hamburger Institut für Sozialforschung, 28 June 1995. 9. See also his The radiation story no one would touch. *Columbia Journalism Rev.* (March/April) 1994, 37–40.
7. Enterline, P. E. Changing attitudes and opinions regarding asbestos and cancer 1934–1965," *Am. J. Industr. Med.* 20:685–700, 1991.
8. Proctor, R. N. *The Nazi War on Cancer*. Princeton: Princeton Univ. Press, 1999.
9. Doll, R. Uncovering the effects of smoking: Historical perspective. *Stat. Methods Med. Res.* 7:87–117, 1998.
10. Proctor, R. N. The anti-tobacco campaign of the Nazis: A little known aspect of public health in Germany, 1933–1945. *BMJ* 313:1450–53, 1996.
11. Proctor, R. N. *Racial Hygiene: Medicine Under the Nazis*. Cambridge: Harvard Univ. Press, 1988. 228, 239–40.
12. Proctor, R. N. The Nazi war on Tobacco: Ideology, evidence, and public health consequences. *Bull. Hist. Med.* 71:435–88, 1997.
13. Mertens, V. E. Zigarettenrauch eine Ursache des Lungenkrebses? (Eine Anregung). *Zeitschr. Krebsforschung* 32:82–91, 1930.
14. Müller, F. H. Tabakmissbrauch und Lungencarcinom. *Zeitsch. Krebsforschung* 49:78, 1939.
15. Schairer, E., and E. Schöniger. Lungenkrebs und Tabakverbrauch. *Zeitschr. Krebsforschung* 54:261–69, 1943.
16. Smith, G. D., S. A. Ströbele, and M. Egger. Smoking and health promotion in Nazi Germany. *J. Epidem. Comm. Health* 48:220, 1994.
17. Macrakis, K. *Surviving the Swastika: Scientific Research in Nazi Germany*. New York: Oxford Univ. Press, 1993. 110–11.
18. Proctor, R. N. Nazi doctors, racial medicine, and human experimentation. In *The Nazi Doctors and the Nuremberg Code*, edited by G. J. Annas and M. A. Grodin. New York: Oxford Univ. Press, 1992. 17–31.
19. Smith, G. D., S. A. Ströbele, and M. Egger. Smoking and death. *BMJ* 310:396, 1995.
20. Lenz, F. *Menschliche Auslese und Rassenhygiene (Eugenik)*, 3rd ed. Munich: J. F. Lehmann Verlag, 1931. 417.
21. Fishkoff, S. They called it mercy killing. *Jerusalem Post (International Ed.)*, 20 April 1996, p. 17.
22. Ramm, R. *Ärztliche Rechts- und Slandeskunde*. Berlin: De Gruyter, 1943.
23. Moll, A. *Ärztliche Ethik: Die Pflichten des Arztes in allen Beziehungen seiner Thätigkeit*. Stuttgart: F. Enke, 1902.

---

[4] I make a similar argument in [34].

24. Grodin, M. A. Historical origins of the Nuremberg Code. In *The Nazi Doctors and the Nuremberg Code*, edited by G. J. Annas and M. A. Grodin. New York: Oxford Univ. Press, 1992. 133–34.

25. Liek, E. *Der Arzt und seine Sendung.* Munich: J. F. Lehmanns Verlag, 1926.

26. Like, E. *Der Welt des Arztes.* Munich: J. F. Lehmanns Verlag, 1933.

27. Klarner, W *Vom Rauchen: Eine Sucht und ihre Bekämpfung.* Nuremberg: Med. Diss., 1940. 43.

28. Steinwallner, B. Zur Pflicht des Heilpraktikers, bei Krebskranken sofort einen Arzt zuzuziehen. *Monatsschr. Krebsbekämpfung* 7:137–38, 1940.

29. Schläger, G. Aufklärung und Verschwiegenheit bei Krebsverdacht. *Monatsschr. Krebsbekämpfung* 10:154, 1943.

30. Niderehe, W. Bericht über die Vorträge betr. Krebsbekämpfung auf dem 3. internat. Kongress f.d. ärztl. Fortbildungswesen in Berlin. 24 Aug. 1937, E 1496 Thür. Landeshauptarchiv Weimar.

31. Hinselmann, H. 14 Jahre Kolposkopie: Ein Rückblick und Ausblick. *Hippokrates* 9:661–67, 1938.

32. Parrish, D. The scientific misconduct definition and falsification of credentials. *Prof. Ethics Report* 9 (Fall):6, 1996.

33. Aly, G., P. Chroust, and C. Pross. *Cleansing the Fatherland: Nazi Medicine and Racial Hygiene.* Baltimore: Johns Hopkins Univ. Press, 1994.

34. Proctor, R. N. Why did the Nazis have the world's most aggressive anti-cancer campaign? *Endeavor* 23: 76–79, 1999. Compare also my *Nazi War on Cancer*, pp. 277–78.

"Ideologically Correct" Science

Reading 8

Michael Gordin, Walter Grunden, Mark Walker, and Zuoyue Wang[1]

Introduction

The historical study of science and ideology is really a twentieth-century phenomenon, for it is only after the First World War that stark differences appeared in political ideology and regime: liberal capitalist democracy, Marxist-Leninist communism, fascism and National Socialism. Furthermore, as the introduction to this volume argues, this historiography has been profoundly influenced by the Cold War, with the result that certain themes have been dominant. This essay will investigate perhaps the most striking examples of science being influenced by ideology, which here will be called "ideologically-correct-science" (ICS). The French Revolution will be included because, as the above-mentioned introduction also describes, it was both the first case study for the interaction of science and ideology, and a surrogate for other Cold War case studies. Not all relevant types of ICS will be, or even could be covered here. Indeed, this article will sacrifice depth in favor of breadth and use the comparative approach in order to provide a suggestive analysis of science under different ideological regimes.

Scholars have generally assumed that a political environment can influence science, but relatively little is known as to how this functioned in particular circumstances or across national boundaries. ICS refers to attempts by the state (or at least some representatives of, or forces within the state) to not only use science, but also to transform it into a more ideologically acceptable form, both with regard to scientific content and institutions. These efforts were often inconsistent, and not always entirely rational, but they existed all the same. Jacobins called for a "democratic," not "aristocratic" science in the French Revolution. Bolsheviks called for a "Marxist," not a "bourgeois" science in the Soviet Union. National Socialists in Germany called for an "Aryan," not a "Jewish" science. Ideologues in Second World War Japan demanded a nationalistic, "Japanese" science and technology. During the McCarthy era in the U.S.A., politicians and some scientists tried to reshape science to help win the Cold War; sometimes calling for what Jessica Wang has described as an "anti-communist" science.[2] Finally, the Red Guards demanded a "people's science" during Mao Zedong's Cultural Revolution in China.

ICS often followed the same pattern: (1) purge of unacceptable scientists and purge or transformation of unacceptable scientific institutions; (2) the enlistment/recruitment of acceptable scientists; (3) the training of new scientists and creation of new institutions; and (4) the production of ICS. This essay will examine this admittedly ideal pattern in order to shed light on ICS in particular and the interaction of science and ideology in general.

In cases of ICS, the state often rewarded scientists and sciences that were, or appeared to be, politically and ideologically correct, while those who deviated from the prescribed path might receive punishment. This pressure was sometimes overt, as during the Chinese Cultural Revolution, or subtle, as in the McCarthy era of the American national security state. ICS could also be self-imposed. Some scientists, voluntarily or otherwise, sought to apply what they viewed as the official ideology in their scientific work. Finally, ICS could also provide "protective coloration,"[3] whereby a straightforward piece of research was wrapped in the official ideology for self-protection or self-promotion. These attempts to make a science that had merely the forms and trappings of ideology should be distinguished from efforts which were made to influence the content, but which fell short and only had an external effect on the social position of the sciences.

Undoubtedly the classic examples of ICS are the "Aryan Physics" (*Deutsche Physik*) movement during the Third Reich and the "Lysenko Affair" in the Soviet Union. Indeed, these are two of the most-studied examples of ideology influencing science. But Aryan Physics, like its counterparts in other sciences under National Socialism, was neither typical of science in the Third Reich, nor very successful. A search for ICS under Hitler would certainly include "Aryan Science," but must also go beyond it. Similarly, Lysenkoism was not typical of science under Stalin. Although it is debatable whether or not it was "successful," it did not spread from plant breeding and genetics to other sciences, and like Aryan science, it did not go unchallenged.

Furthermore, the history of "ideologically-correct-science" is not merely the story of the perversion or destruction of "good" science. Although this was sometimes the case, it is equally true that there are many examples in which ICS either failed to have a particularly harmful effect, or even produced benefits and positive scientific results. This latter case can be made both for a direct effect, as Loren Graham has argued with regard to dialectical materialism sometimes facilitating scientific progress, and an indirect effect, such as Mark Walker's thesis that the fight against Aryan Physics actually strengthened the hands of some in the German physics community.

ICS is a useful concept, and was very real, but it is also something to be used carefully, for it can also be seen as a straw man. If by ICS one means the total, coordinated, systematic, and intentional implementation of an ideologically determined program in science by the state, then ICS never happened, indeed never came close. But interesting and important things did happen. This article will therefore examine several case studies, comparing and contrasting them, with an eye towards gauging the limits and usefulness of this concept.

■ France

The varied fortunes of science in the French Revolution, like the other case studies, is a story of complexity and nuance, which can hardly be given adequate justice here. The two most celebrated incidents of the "attack" on science during the Jacobin Terror—the closing of the Academy of Sciences (*Académie des Sciences*), and the execution of the chemist Antoine Lavoisier—both fall short of being conscious attempts to impose an ideology on science.

Jacobin ideology was only dominant for about a year, and was not synonymous with the French Revolution. Moreover, all sciences were not treated equally: important Jacobins were hostile to abstract, theoretical, and mathematical science, but were favorable to natural history. After much debate about how to reform education and how to structure expertise in the Republic of Virtue, in 1793 the Academy was closed as an institution and its resources dispersed. But the issues here were not just science, and even not centrally science. Resentment of royal privilege and corporate prestige were mainly responsible for the closing of the Academy.[4] While the Academy was eliminated as a vestige of royalist corporatist elitism, the Jardin du Roi was kept intact.[5] Moreover, Academy scientists (for the most part) were subsequently employed by the government of Revolutionary France.

The case of Lavoisier, on the surface the purge of an unacceptable scientist who was hardly a strong advocate of Robespierre's government, upon inspection had even less to do with science than the closure of the Academy. Lavoisier was executed by the machinery of the Terror, but not for any reason connected to his positions in science. His association with the Tax Farm sealed his fate. Similarly, Condorcet died at the hands of the Revolution, not because of his mathematics, rather because of hostility to his rationalist Enlightenment views.[6]

The Terror prematurely ended the lives of several prominent scientists, and interrupted the careers of many others, forcing twenty academicians out of forty-eight into "exile" (most went to the provinces, and only about four actually emigrated during the Jacobin Republic). But other natural scientists in educational institutions, government branches, and other venues were actively recruited by the regime. In the case of the Jardin des Plantes an old royal corporate research institution was maintained, albeit with profound alterations (see below). During the period of revolutionary wars, members of the Academy of Sciences received government contracts to fulfill old academic projects like the metric system, gunpowder production, and military engineering. Most aided the Revolution with little grumbling, or even with little attention to other political events once the guillotine lost its central prominence.[7] In contrast, some scientists, like the chemist A. Fourcroy (1755–1809), not only did not resent the Jacobins, they were active members of the state administration.

During Napoleon Bonaparte's Empire, the state recruited scientists even more avidly than before, and many scientists were quite eager to serve Napoleon's "technocratic" regime and appreciated the return to stability he represented in their eyes. The state now poured an enormous amount of money into military institutions, and scientific and technical expertise were richly rewarded in new institutions like the École Polytechnique, which was founded under the Directory and blossomed under Napoleon.[8] Given the prominence of scientists in the advisory apparatus of the Old Regime, especially in the form of the Academy of Sciences, this development under the Directory and the Empire can be seen as a continuation of previously established practices. The persecution of "unacceptable" scientists seems more the aberration from French practice, and the recruitment of those old specialists who were acceptable continued as before, albeit in new institutions.

Since it is difficult to discern a clear ideology in the French case governing the selection of acceptable as opposed to unacceptable scientists, it is also difficult to find an ideological criterion for training new scientists. After the ideological excesses that led to the closure of the Academy had passed and some of the dust of the Terror had settled, France still faced a series of foreign wars, and sorely needed technical expertise and noticeably lacked qualified young practitioners.

The solution was the creation of a new system of Grandes Écoles, headed by the École Polytechnique, designed to inculcate military discipline into young minds drawn from all corners of France with the necessary technical and mathematical skills required by the emerging modern bureaucratic state and a conscripted army.[9] The plans for such a system of education originated in the Old Regime, whereby places in military schools would be attained through nationwide competition and nominations by local authorities. The subject matter of all schools, especially the Polytechnique, was heavily imbued with mathematics applied to concrete problems of military necessity.

The results were impressive: a new generation was trained, filling the officer ranks of the army, the upper levels of the bureaucracy, and other positions at the top of the modem nation-state. The ideology of discipline through mathematical scholarship and service to the State has been so well inculcated into French society that it is scarcely noticed how large an influence this, the most ideological of Napoleonic institutions, still plays in modern France.

The attack on Newtonianism and the retention of the Jardin du Roi as the ardin des Plantes after the dissolution of the Academy of Sciences are the two nost prominent candidates for ideology affecting science. Charles Gillispie and L. Pearce Williams have made the most forceful case for an anti-Newtonianism, arguing that an ideology of anti-elitist Romanticism, deriving from Rousseau and Diderot, motivated the Jacobins' hostility to an atomized, mathematized, Newtonian universe.[10] Many of the individuals in question expressed hostility to some aspects of the Newtonian worldview, and specific examples of ideology did seep into the way debates about the content of science were conducted.[11] Similarly, because of the French Revolution, an exact, scientific biology was displaced by a discursive descriptive natural history within the Jardin des Plantes.[12]

But these examples of ideology and science do not cut to the actual core of the politics of science during the French Revolution. It is unclear how much of a case can be made for the application of specific Rousseauvist anti-Newtonian ideals during the height of the Terror. Even the leading anti-Newtonian ideologue of the time, Marat, was motivated less by an ideological stance towards Newton than by revenge against the Académie des Sciences for its treatment of him during an earlier scientific dispute.[13] Perhaps more startling is the fact that these trends were not followed up once the Academy was closed and Lavoisier had been executed. Laplace wrote his great Newtonian masterpiece after the debates recorded by Williams and Gillispie, for example. But the fact that the attempt to reject or replace Newtonianism did not succeed does not mean that it should be dismissed.

The reformation of the Jardin du Roi into the Jardin des Plantes was more successful during this period. It is noteworthy that, despite the rhetoric of anti-corporatism that led to the demise of the Academy, the Jardin remained well financed and continued to perform the same sort of botanical research as it had under the Old Regime.[14] The secret of this persistence was the articulation by the botanists of a natural historical program they favored that also seemed to accord with the hostility to abstraction noted by Williams and Gillispie. As a result, this kind of empirical botany was encouraged. A particular type of ideologically-correct science was indeed adopted as a form of basic research, but basic research led the way here in defining what it meant to be ideologically correct.

ICS during the French Revolution was fleeting and superficial. At the same time that the Reign of Terror reached its height in Paris and de-Christianization was ravaging the French countryside, most scientists, who were products of the Old Regime, were put to work for the Revolution; others needed only to repackage their research in an ideologically congenial wrapper. The scientific community did have to make concessions to the new political order, and a few individual scientists (for example, Condorcet and Lavoisier) suffered and died, but in general French science and scientists benefited from the French Revolution and Napoleon. The push for ICS did not produce a significant change in the content of research. However, ICS and the greater political and ideological currents of the French Revolution did create new scientific institutions and influence research programs and thereby led the way in terms of political control of scientists and scientists' accommodation.

■ The Soviet Union

The reconstruction of Russian science was well on its way to a socially more useful institution before 1917, but the chaos of the First World War and the transformation of Russian society by the Bolshevik Revolution also brought profound change to Russian science. Perhaps surprisingly in a nation where science had never penetrated much farther than a vanishingly small percentage of the literate intelligentsia, the Bolsheviks and their allied parties saw reform of higher education, and therefore indirectly, science as one of the first tasks on their agenda. At almost the first opportunity they approached the Imperial Academy of Sciences in Petrograd with both carrot and stick in hand.[15] On the one hand, the largely bourgeois membership of the Academy was reluctant to extend a badly needed hand to the new regime's quest for rapid economic improvement and technical expertise. On the other hand, compelling reasons kept the vast majority of Russian scientists from emigrating and induced them to cooperate, however reluctantly at first, with the regime.

It was by no means certain that the regime would last, and there seemed to be every hope of the regime moderating its rhetoric against bourgeois specialists (as indeed it did). Such hope gave those with a sincere patriotic bent an opportunity to actually put some of their practical suggestions to work. The thrust of the early period of science policy in the Soviet Union was more the establishment of new scientific institutions rather than the destruction of older ones. Moreover, the Bolsheviks offered unheard-of blandishments in the form of prestige, equipment, and funding in an effort to persuade the academicians to lend some assistance.[16] Finally, the Bolshevik regime seemed the lesser of several evils, as rabidly "anti-specialist" movements like *Proletkul't* hovered on the horizon.

During the period of War Communism (1918–21), *Proletkul't* was a vibrant cultural movement, one that marked the first effort to establish ideologically-correct-science in the Soviet context. *Proletkul't* was by no means directed exclusively, or even principally, towards science. It argued that a proletarian state required a proletarian culture, not the realism of capitalist art, the individualism of capitalist literature, or the technocracy of capitalist science.[17] Technocratic specialists and other remnants of tsarist capitalist culture would have to go, and a more democratic and proletarian science would be imposed in its stead.[18]

Thus for quite some time, bourgeois specialists did not know how they would be treated by the Bolsheviks, especially since there was now a "Communist Academy" alongside the traditional Academy of Sciences. But Lenin had little patience for such efforts to alienate much needed specialists, and his eventual suppression of *Proletkul't* and the closing of the Communist Academy served as a signal to bourgeois scientists that their kind would be tolerated as long as they were amenable to the new regime's demands.[19] In fact, except for an exile of some 200 dissident intellectuals, there was not much of a purge in science until the period 1928–31.

What it meant to be an "acceptable" scientist in the early Soviet Union fluctuated widely with the attitudes and needs of the fledgling regime. During the period of War Communism, bourgeois specialists were (officially) "unacceptable," but were nevertheless used. As the Soviet economy began to falter, the regime began to accommodate those in possession of needed technical skills. The heyday of the bourgeois specialist—the period of the New Economic Policy (NEP, 1921–27)—induced many scientists who were ideologically opposed to Bolshevism to make a temporary peace, while giving the Soviet regime time to lick its fiscal wounds and gear up for socialist industrialization.

The essence of the NEP attitude towards bourgeois scientists and engineers had already been expressed by Lenin in his opposition to the iconoclastic fury of *Proletkul't*. Technical expertise would always be necessary, and as long as those who had the knowledge would only share it if given sufficiently high salaries and ideological breathing room, then they should be afforded those luxuries.[20] This did not mean, however, that ideological constraints were put on hold. Ideologically "acceptable" scientists, like the young Lev D. Landau, for example, were actively encouraged by the regime and promoted over old bourgeois specialists who staffed the old universities.[21] Indeed, the communists made a fundamental distinction between research institutes where scientific ability was most important, even if not accompanied by appropriate political conviction, and the universities, where only politically reliable scientists would be used to train young scientists. It is interesting to note that National Socialists in Germany and to a certain degree the Communist Party in China made the same distinction.

For the time being, the Soviet state only mildly harassed those who chose to hold to their old views—provided their skills were truly indispensable to the industrialization of the new regime. The Shakhty trial of 1928 changed all this, however, when bourgeois specialists were accused of "wrecking" and industrial sabotage designed to cripple Soviet power.[22] The honeymoon had ended, and under the rising power of Joseph Stalin, bourgeois specialists were not tolerated during the years 1928–31. Thereafter the term "bourgeois specialist" was no longer used, and many former "specialists" quietly returned to the positions of prominence in science they had enjoyed before. The Soviet nuclear weapons project, which used both former specialists and younger scientists trained under the Soviet educational system, was typical in this regard.[23]

The generation of new cadres of ideologically suitable scientists and technicians constituted one of the most important aspects of early Soviet science policy. The splitting of research and education was the first stage in this development. Education was placed entirely in the hands of ideologically sanitized pedagogues within the People's Commissariat of Popular Enlightenment (*Narkompros*). Research institutes were left under various economic Commissariats, and permitted a more eclectic personnel. At the same time, the State transformed the research institutes, borrowing some aspects of Western organization for individual labs, but placing them all into a Soviet framework.[24]

While some bourgeois specialists were prominent in this framework, by the early 1930s and the conclusion of the first of Stalin's Five-Year Plans, most had either blended into the woodwork and adopted appropriate ideological colors or had been executed or exiled during the purges.[25] However, as late as the 1950s, non-Party scientists, former "bourgeois specialists," occupied the majority of high-level administrative positions in Soviet scientific research.

The Soviet Union grudgingly used its "bourgeois specialists" while simultaneously training new cadres of "red" scientists. But these new ideologically correct scientists had the same professional aspirations as their "bourgeois" mentors—concern for international scientific standards, the need for international contacts—and therefore sometimes clashed just as forcefully with the Stalinist regime's desire for ideologically fidelity.

The growth of cadres of communist researchers in the various fields of science was really quite extraordinary.[26] As more and more competent (and ideologically "clean") individuals were generated, they began to take over positions held by disgraced members of the older generation.[27] By the onset of the Second World War, Stalinists had essentially completed the ideological purification and installation of "red" specialists in almost all levels of the Soviet research empire.

Dialectical materialism, the official philosophy of science of the Soviet Union, complemented historical materialism—the Marxist theory of historical and economic development—to compose the complete orthodox set of beliefs about the social and natural world. The nature of dialectical materialism and the extent of its epistemological and ontological grasp had been a source of debate since the early interpretations of Engels' and Lenin's writings on the natural world during the 1920s.[28] But unlike the meaning of dialectical materialism, the unwritten "requirement" that scientists hold to some of its tenets (or at least not openly contradict them) was more or less constant through Soviet history—and reached some disastrous consequences during the Lysenko years. Yet the historian Loren Graham has pointed to another side of dialectical materialism, which he calls the "authentic phase."[29] Graham argues that dialectical materialism was sometimes used by scientists freely as a positive force for scientific reasoning.[30] Graham's persuasive argument thus makes the intriguing point that scientists often adopted ICS as basic research in the Soviet Union voluntarily (or semi-voluntarily) and occasionally used it to produce significant results.

Whereas dialectical materialism provides the most important example of ICS is basic research in the Soviet Union, the most famous instance of ICS as applied research is the well-known case of Lysenkoism.[31] Scion of a Ukrainian peasant family, Trofim Denisovich Lysenko began his work as an agronomist in the 1920s in an agricultural station near Baku. While there, he claimed to have discovered a biological process he dubbed "vernalization" (*iarovizatsiia*): the treatment of germinated seeds of various plants with abnormal conditions of eat, cold, and other forms of environmental exposure, in order to make plants develop in a more appropriate way—essentially a neo-Lamarckian biological program. Lysenko's attempts to present his results to the Soviet agronomic and genetics community were rebuffed as contrary to all known facts about genetics. The famed geneticist N.I. Vavilov initially supported Lysenko's research as potentially producing innovations in agronomic practice, but broke with Lysenkc when he started to push his neo-Lamarckian views on plant breeding and genetics in the mid-1930s.

In the early 1930s, Lysenko teamed up with ideologist I.I. Prezent, who convinced Lysenko to link his neo-Lamarckian views of inheritance with Darwinism, and to couch both in a Marxist framework. This marriage of dialectical materialism and agronomic practice in opposition to genetic theory caught the attention of Stalin in the late 1930s, who praised Lysenko openly in various contexts as a means of supporting the regime's disastrous and bloody collectivization campaign in the countryside. Lysenko grew in power, becoming president of the Lenin All-Union Academy of Agricultural Sciences (VASKENIL) while its former president, Vavilov, was arrested on grounds of counter-revolutionary and anti-Soviet activity, and died of malnutrition shortly before being released from prison in the early 1940s. The actual banning of genetics did not happen until after the Second World War, when Lysenko's star actually seemed to be waning and the Cold War got underway. But in 1948, Lysenko read a speech (toned down but supported by the personal editing of Stalin)[32] condemning genetics as a "bourgeois" science and banning almost all research on it in the Soviet Union.

While Nikita Khrushchev liberalized much of the terror apparatus of Stalin's state after the latter's death in 1953 and especially after the Twentieth Party Congress in 1956, he liked Lysenko personally and continued to fund him lavishly and support a series of disastrous agricultural programs which his favorite proposed. After Khrushchev's fall in October 1964, however, Lysenko's days in power were numbered. Genetics was restored in 1965, but the recovery process was painful and the loss of Vavilov hard to forgive. The scars caused by Lysenkoism remain to this day.

Lysenkoism, which influenced Soviet science for decades, was obviously ideologically-correct-science. But it was eventually overthrown by forces within Soviet science and society, and the science it had discredited and dismantled, genetics, was reinstated and rebuilt. Lysenko and his followers also failed to extend their influence to other Soviet sciences. In particular, physics was able to rebuff Lysenko-inspired attacks on certain aspects of modern physics, both because of the relevance and irrelevance of physics to Stalin's foreign and domestic policies. At first physics was not very important to the Soviet leadership, for in contrast to Lysenko's theories, it promised neither to solve the country's problems, nor fit particularly well into Soviet ideology. When the Second World War began and the potential of nuclear weapons was clear, physics became far too important to purge or distort. Thus Lysenkoism was arguably exceptional, and reveals little about the other major purges of Soviet intellectuals that were attempted in the period after the Second World War.[33]

■ Germany

During the first year of National Socialist rule in Germany, a significant percentage of scientists (perhaps as much as 15 percent) were forced out on racial and political grounds.[34] This purge was not aimed particularly at scientists or science—the campaign against Albert Einstein is the exception that proves the rule—but rather was a consequence of the National Socialist "cleansing" of the entire civil service. This larger purge was itself apparently a largely unplanned, if not spontaneous reaction to the failure of the nationwide boycott of Jewish businesses in April 1933. The effect on science was tremendous, but this purge does not demonstrate any plan or intention on the part of the National Socialist leadership to create an Aryan science.

There is evidence for more direct interest by Hitler's government in the transformation of scientific institutions, but again a close look reveals different priorities. The universities were purged and transformed right away because they were educational institutions charged with the training of German youth.[35] Their transformation was profound, but also hidden. The structure of the university remained largely intact, at least on paper, but most autonomy was robbed by the introduction of the "leadership principle": a strict hierarchy, whereby one had to obey everyone above, but could order about everyone below. In principle, the faculty still met and voted, prepared lists of candidates for positions, and so on, but in practice the deans and rector— political appointees, of course—often had almost dictatorial power. Other scientific institutions, including the Kaiser Wilhelm Society and its research institutes, as well as the various Academies of Science, were transformed in a similar way, although significantly later in the Third Reich.[36]

The dates by which all Jewish members had been purged provides one of the most telling indications of how relatively unimportant these research institutes and academies were for the National Socialist leadership. Whereas most Jewish scientists in the universities had lost their positions in 1933, other Jewish scientists managed to remain at research institutions for many years. The Prussian Academy of Sciences, for example, under pressure from the Ministry of Education, finally asked the last Jewish member to resign in 1938, shortly before *Reichskrystallnacht,* the nationwide orgy of violence against German Jews. In 1933 the Education Ministry had been keen to publicize its treatment of Einstein. After he had already resigned from the Prussian Academy, the Ministry pressured its leadership to issue a press release, essentially saying good riddance. But in 1938 the Ministry wanted to keep the final purge of Jewish scientists quiet, so as not to publicize the fact that there were still Jews in the Prussian Academy.[37]

The purge of German science by the National Socialists makes clear how little interest Hitler and his followers had in scientific research. Their interest was aroused only when scientists demonstrated that modern science could serve National Socialism. It is striking and depressing to see how quickly the vacancies caused by the purge were filled by generally competent, racially and politically acceptable scientists eager to serve the new regime. For those who were already in place, it usually sufficed to demonstrate Aryan status and an apolitical attitude to keep their jobs, although they were pressured to yield greater political and ideological cooperation with the National Socialist movement.

Those moving up—i.e., who did not already have permanent positions—had to be both Aryan and willing participants in the political and ideological rituals introduced by the National Socialists into the universities. Such rituals included or encompassed attendance at "political" indoctrination camps, membership in National Socialist organizations, including, of course, the National Socialist German Workers Party (NSDAP), and other forms of participation in National Socialism. The authorities paid little attention to their research when it came to judging political acceptability, although this might well enter into whether or not they were hired.

Rearmament, especially beginning in 1936, offered great opportunities to scientists who had something to offer the regime; similarly, the new racial hygiene policies (sterilization, "euthanasia," restrictions on marriage) provided great opportunities for physicians, biologists, anthropologists, psychologists, and psychiatrists.[38] In general, most scientists did not make the transition to the racist, Aryan science, but did adapt themselves and their research in order to work under National Socialism.

Perhaps the greatest failure of National Socialist science policy concerned the training of the next generation of scientists. The politicization of education and emergence of National Socialist youth organizations like the Hitler Youth eroded both the quality and the quantity of scientific education. The creation of new institutions was not much more successful. The National Socialists did not really try very hard to create new scientific institutions. Even the *Ahnenerbe* ("Ancestral Heritage"), the scientific research arm of the SS, relied mainly on research contracts in order to encourage certain types of scientific work.[39] Most truly new research institutions created by the National Socialists had little to do with science, and were so ideological and politicized that they were really incapable of producing significant basic or applied research. Scientific work was done for the National Socialists, but it usually took place in institutions Hitler's movement had inherited. The rocket project is the exception that proves the rule: Army officers and engineers, not National Socialists, created and developed it during the Third Reich.

The politicization of the universities was compounded by the carnage of war, as very many students were offered up as cannon fodder during the Second World War. The result was a generation lost to German science: with few exceptions, only scientists who had entered the university system during the Weimar Republic survived the Third Reich. Despite the regime's attitudes towards women, by the middle of the war women made up a large percentage of university students because so many of their male counterparts were fighting and dying on the front.[40]

The regime wanted the help of scientists and physicians in order to provide a scientific basis for their racist, and eventually murderous race hygiene. But despite the active participation of scientists and physicians in this program, and the infamous experiments carried out at Auschwitz and elsewhere on unwilling concentration camp inmates,[41] researchers could not deliver scientific proof of the supremacy of the Aryan race. In retrospect, this was revealed by the infamous Nuremberg Laws in 1935. When the National Socialist State finally issued the binding legal definition of what "non-Aryan" meant, it had to fall back on to a religious, not racial definition of who was a Jew.

However, there were also scientists who claimed to be practicing Aryan science when carrying out their basic research. These were not race hygienists—although a biologist claimed to be fighting for an "Aryan Biology" (*Deutsche Biologie*)—but rather physicists, mathematicians, psychologists, chemists, and engineers. These Aryan movements in German science and technology[42] eventually failed in their efforts to seize control of their disciplines precisely because they were barren in the National Socialist sense. The rulers of Germany wanted science useful to them, and it was the leaders of the established scientific communities, not the Aryan scientists, who were able to gain and retain the backing of influential patrons in the National Socialist state.

Thus the calls by the Nobel laureates and "Aryan Physicists" Philipp Lenard and Johannes Stark to eliminate the influence of Jews in physics, and the "Aryan Mathematician" Ludwig Bieberbach's assertion that Aryans and Jews made different types of mathematics eventually fell on deaf ears, while the applied mathematician Ludwig Prandtl and the theoretical physicist Werner Heisenberg offered their expertise in designing wind tunnels and nuclear weapons, respectively. Established scientists like Heisenberg and Prandtl were thereby able to sideline or neutralize their "Aryan" colleagues by convincing political leaders that their basic and applied research might facilitate both military conquest by creating new and improved weapons[43] as well as the racial engineering of Europe by providing new methods for distinguishing Aryans and non-Aryans.

Ideologically-correct-science had a profound effect on German science, far beyond the well-known case of the ideological attacks by Aryan Physics on modern physics. But the end effect was almost always defeat for the ideologues, is the established scientific communities sought refuge and support from leading National Socialists. The main effect of "Aryan Science" was to drive most other scientists further and faster down into the arms of National Socialism, making themselves more useful and relevant for the often murderous policies of the regime.

The very anti-intellectual climate within the ranks of leading National socialists worked against the advocates of Aryan Science. A National Socialist Ideologue disdainful of science also had little interest in its Aryan variant, while the technocrats scattered throughout the National Socialist hierarchy naturally threw their support behind established scientists who could deliver the goods.

Scientists helped build rockets[44] and jet planes,[45] researched new biological and chemical weapons (which fortunately were not used), and dangled the prospect of "Wonder Weapons" like the atom bomb[46] before leading National Socialists. Researchers ranging from physicians, biologists, psychiatrists, psychologists, anthropologists, economists and geographers[47] helped implement the murderous race hygiene policies of "euthanasia" and "Germanization," and finally murder and genocide by helping to select the victims and create new and more effective ways to torture and kill them.

■ Japan

During the Second World War in Japan there was no need for a racial or ethnic cleansing in the sciences because, with the exception of a few notable Koreans, all scientists were Japanese and shared essentially the same racial, ethnic, and cultural identity. Moreover, because of the relative paucity of scientists with advanced scientific training in Japan, even Koreans—whose homeland Japan had occupied before the war and who were considered second-class citizens in Japan—were allowed to retain their positions at the university.[48]

There were also few incidents in which scientists were jailed for expressing anti-imperialist views. In one case, several members of an academic research group, including the physicist Taketani Mitsuo, were arrested for advocating resistance to Japanese imperialism through their serial publication, *Sekai bunka* (World Culture). Taketani was detained—allegedly for his research activities on natural dialectics—accused of helping to promote the Communist Party in Japan, and forced to state that he had acted under instructions from the Comintern. The judge who reviewed his case, however, suspended prosecution and released him to the custody of his colleague and close friend, Yukawa Hideki.[49] On the whole, such instances were rare.

Under military influence, the government enacted numerous laws in the 1930s to acquire greater control over the people and the economy. As part of the militarization of the nation, institutions of scientific research, such as the imperial universities and the prestigious Institute for Physical and Chemical Research (also known by its Japanese acronym "Riken"), were also brought under the aegis of the military. By influencing budget allotments for basic and applied research, as well as the production orders for the resulting manufactured items, the military began to have a significant impact on scientific research. As a result, there was no need to transform such institutions; only greater administrative and economic control proved necessary.[50]

Nevertheless, the recruitment of first-rate scientists became a critical problem for the military. Nearly all of the nation's most famous and competent scientists had spent years abroad studying in Western nations, and they were thus considered suspect by the military leadership, which was by and large xenophobic in its worldview. Moreover, most scientists at the university had little interest in suspending their own research for military projects.[51] Out of necessity, however, the military consulted leading scientists at the universities when those in uniform—who were usually little more than higher-school educated technicians—proved incapable of advanced level research. Both the army and navy, for example, turned to physicist Nishina Yoshio, Director of the Physics Department at the Riken, to complete a feasibility study of the possible exploitation of nuclear energy for military purposes.[52]

Like Nazi Germany, Japan proved terribly shortsighted in preparing the nation's scientific infrastructure for a prolonged and total war. One particularly stable area of failure was in the training of new scientists and the creation of new institutions of science and technology. The military continued to draft students from university science programs and departments throughout the war. This was halted only through the concerted efforts of senior scientists like Nishina, who insisted upon military deferments for designated students of exceptional ability in exchange for agreeing to conduct research for the military. Only by such means was the older generation able to preserve the next generation of scientists.[53]

As for new institutions of science, by the early 1940s, the resources for their construction and maintenance were dwindling and the move was toward consolidation and rationalization, not expansion or the creation of new scientific institutions. The capstone of the trend came in early 1942 with the establishment of the Board of Technology. Roughly analogous to the Office of Scientific Research and Development in the United States, the purpose of the Board of Technology was to coordinate scientific research and development of new technologies between civilian and military institutions, as well as between the army and navy.

The organization looked impressive on paper, but in reality, the Board of Technology proved a dismal failure. It never acquired sufficient authority or capability to supersede the numerous administrative boundaries that such a task entailed. It could not overcome the substantial compartmentalization of civilian and military research. Neither could it redirect the complex network of financial arrangements, production contracts, and social ties that each military service had to its preferred *zaibatsu* (industrial combines) and university cliques. The board had little success in overcoming the bitter enmity that existed between the army and navy to convince them to collaborate on key projects until the war was already all but lost.[54]

There was no single ideology in wartime Japan comparable to Aryan Physics a Nazi Germany, nor was there any ideological movement in the sciences that gained such comprehensive state support and promotion as Lysenkoism did in the Soviet Union. Beyond the ubiquitous rhetoric of national militarism that emphasized sacrifice and service to the Emperor and nation, there was no prevailing ideology to impact science as there was in Germany or the Soviet Union. Yet there was a call for a distinctly Japanese form of technological development based on the nation's situational imperatives, that is, the rise in demand for military production in the face of rapidly diminishing raw materials.

Japan's progress in science and technology had been dependent upon the cooperation of Western nations and foreign teachers since the late nineteenth century, but as the military leaders of Japan drove the nation inexorably toward world war in the 1930s, increased hostility toward the West compelled bureaucrats and intellectuals alike to question the West as a model for Japan's technological development. Having no indigenous model as a substitute, a distinction had to be drawn between American and German paths of development. The rationalized "German path," it was argued, was better suited to Japan, as it aimed to limit the use of raw materials and promoted the use of substitutes. When its German ally proved niggardly in technology transfers, however, Japan was forced again to look inward.

Imitation of the West was to be rejected in favor of a uniquely Japanese path of technological development in accord with the nation's paucity of natural resources.[55] As one ideologue stated:

> The resources of the Greater East Asia Co-Prosperity Sphere are awaiting the creation of the new technologies that will make most effective use of them. It is only then that these resources will acquire value. The existence of scientific research, which may give birth to this new technological creativity, will provide a firm basis for the cultivation of the Co-Prosperity Sphere, and for this reason the promotion of such research is currently an urgent necessity.[56]

Despite the flurry of mobilization and rationalization laws and measures that were enacted to realize this vision, no distinctively "Japanese" science emerged, nor was Japan able to free itself of its pattern of technological borrowing and dependence upon the West. Toward the end of the war, when necessity and desperation drove the nation's leadership to extremes, the government and military called upon scientists and engineers to draw inspiration from Japan's traditional past and the unique characteristics of the Japanese people, all in the effort to create an ideological rallying point for Japanese scientists in the development of some new weapons technology that could turn the tide of war in their favor.[57]

By 1943 the National Socialist state was also calling upon its scientists, engineers, and even inventors to create "wonder weapons" which would use qualitative superiority to overcome the quantitative superiority of its opponents. The Japanese military's answer, however, was no miracle weapon of science, such as a rocket or an atomic bomb. Rather, it responded with crude suicide craft, such as the *Ohka* piloted missile and the *Kaiten* midget submarine. No "Japanese" style science emerged from the war, and the ultimate "Japanese path" of technological development resulted in death for many of the nation's youth.

In the case of Japan, there was no readily identifiable ideologically-correct-science. Rather there was only a vague policy objective to guide technological development that was in accord with the nation's situational imperatives. The lack of natural resources, coupled with the nation's Spartan industrial infrastructure and its limited scientific and technological capacity, predestined Japan's fate in a total war against the United States. Japan proved incapable of producing such wonder weapons as the atomic bomb, long-range guided missiles, and advanced radar, but its army did, for a time, strongly support the development of biological weapons and made significant advances in this field.[58]

Biological weapons were easy to mass-produce with few resources, and thus fit within the military's vision of a new weapon derived from a "Japanese technological path." Ultimately, however, the fear of a response in kind from the United States and the ever-present possibility of a boomerang effect appear to have deterred Japan's use of biological weapons on a wider scale beyond the war in China. As a result, biological weapons never became a significant factor in determining the outcome of the Pacific War, and the military never acquired a uniquely Japanese wonder weapon of his own.

■ United States

If communism served as a powerful ideology of science in the East, especially in the Soviet Union and China, did anti-communism play the same role in the West during the Cold War? Since American McCarthyism in the late 1940s and 1950s represented the peak of this political and cultural phenomenon, we might expect to see signs of the search for an ideologically correct science.[59] However, there were very few unambiguous examples of efforts to influence scientific content that were motivated or constrained by anti-communism as a political ideology. Rather, what does emerge clearly is the pervasive influence of anti-communism on the *political* roles of scientists whose professional identity assumed significant but not overwhelming importance.

Scholars now generally agree that the two characteristics of McCarthyism, domestic anti-communism and the denial of due process to those accused of communism, existed both before and after the period when Senator Joseph McCarthy made the cause his personal crusade in 1950—4.[60] The Cold War ideology of anti-communism not only saw a direct threat to American security in potential Soviet expansion on the international front, but also from perceived communist subversion at the domestic front. The national security state organized national life around national defense and fed on the Cold War ideology of anti-communism. It dominated science and technology policy, and thus indirectly but powerfully shaped American scientists' political and scientific activities.

On the one hand, a large number of American scientists engaged directly in the making and testing of nuclear weapons in the national laboratories of the Atomic Energy Commission (AEC). Those in academic and industrial settings also came to depend on the defense establishment for funding. As Paul Forman has argued, this dependence tended to make scientists choose, consciously or unconsciously, research directions that would benefit their patrons.[61] On the other hand, scientists were persecuted during the McCarthy era because their past association, political opinions, or policy advice deviated from the political orthodoxy prevailing at the time.

The security clearance case of J. Robert Oppenheimer was perhaps the best-known example of McCarthyist attacks on scientists. As the famous director of the Los Alamos laboratory that created the atomic bomb during Second World War, Oppenheimer was nevertheless stripped of his security clearance in 1954 by the Atomic Energy Commission (AEC). The AEC decision cited past association with radical causes, opposition to the hydrogen bomb, and "defects in character."[62] Hundreds of other, less-well known scientists suffered similar or worse treatment both before and after the Oppenheimer case.[63] The U.S. State Department denied passports to a number of American scientists with liberal reputations so they could not travel abroad. It also refused to issue visas to some foreign scientists who wanted to visit the U.S.A.[64]

The Oppenheimer case evoked a most vehement protest from the scientific community, which generally blamed the injustice on a paranoid security system.[65] It undoubtedly marked a profound deterioration in the relationship between many American scientists and the national security state. The far-reaching repercussions did not escape the top government officials. Eisenhower, while agreeing with the AEC's decision, nevertheless worried about the case's effect on scientists in various defense projects. Aware of the potentially explosive impact on scientists and dangerous exploitation by McCarthy, Eisenhower told his aides that "we've got to handle this [Oppenheimer case] so that all our scientists are not made to be Reds."[66] Eisenhower wrote to New York writer Robert Sherwood shortly before the AEC decision, stating that because he was "so acutely conscious of the great contributions the scientists of our country have made to our security and welfare," he shared the hope that Oppenheimer could be cleared.[67]

It was to the President's and the AEC's relief that the "mass exodus" from weapons laboratories, as predicted by various scientific groups, failed to materialize in the wake of the Oppenheimer case.[68] While Eisenhower may have feared losing the services of scientists, there was no lasting damage to the Cold War partnership between science and the State. Scientists had warned since the late 1940s that unfair security procedures would lead scientists to desert government positions, but the threat was more rhetoric than reality.[69]

The fact that few scientists left their government research positions in the aftermath of the Oppenheimer case indicates that scientists, especially younger scientists, learned to live with the new Cold War political economy of science, what Eisenhower would call the "military-industrial complex." As physicist Herbert York, first director of the AEC's Livermore Laboratory on nuclear weapons in the 1950s, later reflected, young scientists had the practical needs of finding jobs that matched their training and supported their families.[70] Others simply recognized the need to combat two "Joes," both Joe McCarthy and Joe Stalin, especially after the outbreak of the Korean War.[71] In any case, since the newly-founded National Science Foundation only slowly gained substantial budgets, the funding structure of American science was skewed toward the military, which left few alternative sources of financial support to scientists, including the training of new ones.[72] Consciously or unconsciously, scientists were integrated into the national security state.

In the United States during the McCarthy era, there was a general acceptance, either tacit or explicit, by both individuals and institutions, of loyalty oaths and security clearances. Most universities, for example, refused to hire known communists. In 1950 the regents of the University of California adopted a requirement that all university employees had to sign a loyalty oath stating that they were not members of the Communist Party. As a result, dozens of faculty members, many of them scientists, left for other institutions in protest while some others, including tenured professors, were fired from the university.[73]

The same year, scientific organizations such as the Federation of American Scientists and the National Academy of Sciences fought successfully to remove amendments in the National Science Foundation Bill which would have required applicants to the foundation for unclassified research to undergo security clearance and an FBI background check. But, with dismay, they felt that they had to accept the requirement of loyalty oath in the bill as a compromise to get it passed.[74] There were a few nuclear scientists who deliberately switched to fields where they did not need security clearance. Leo Szilard, for example, turned to molecular biology.[75] Others, like Philip Morrison, continued in their fields but avoided work that would have required them to apply for a security clearance.[76]

Without belittling the pains and the injustice that the victims of Cold War anti-communism suffered, a distinction should be made between Stalinist/Maoist communism and American McCarthyism. Few American scientists were persecuted through State-sponsored violence for their particular beliefs in science, in contrast to what happened in Stalinist Russia and Maoist China, although they were repressed in other ways. While Vavilov starved to death in Stalin's prison for resisting Lysenkoist theory and while dozens of prominent

Chinese scientists were killed during the Cultural Revolution for being bourgeois "reactionary academic authorities" (see below), the worst that happened to American scientist-victims of anti-communism was, with few exceptions, that they lost their jobs, or as in Oppenheimer's case, their security clearances.

The consequences of McCarthyism went beyond the harm done to individuals, however. In fighting dictatorial communism, American anti-communist crusaders adopted the same anti-democratic tactics employed by the enemy. The American left, including that in science, was largely silenced and social reforms aborted. Fearing the charge of "being soft on communism," Cold War liberals energetically led the U.S.A. into a costly and misguided war in Vietnam.[77] Few of those American scientists who served as major advisors to the government in the 1950s and 1960s, for example, represented the liberal and left-wing positions briefly influential in the immediate post-Second World War years.[78] The significance of McCarthyism "may well have been in what did not happen rather than in what did."[79] In science, for example, we will never know what scientific research could have been pursued during the Cold War had there not been pressure to work in the national security state system. Likewise, we can only speculate on whether scientific research more closely related to civilian technology could have been advanced further and earlier than it did.

Scientists, especially nuclear scientists associated with government labs, were certainly subject to special scrutiny from the government because of their perceived access to "atomic secrets." Yet, the ideological impact of anti-communism on American science was only general and indirect. In contrast to Marxism, which attempted to function in Stalinist Russia and Maoist China as an all-encompassing ideology with specific doctrines governing science and philosophy, American anti-communism was primarily a political ideology. It guided American foreign policy and influenced domestic politics, but never set down a number of doctrines to be followed in science and philosophy.

The excesses of the McCarthy era in the United States had a profound effect on scientists and their relationship with the state, but it did not lead directly to a new type of an ideologically-correct "anti-communist" science like Aryan Physics or Lysenkoist biology. In the United States, ICS manifested itself as the connection between research and the goals of the State, in particular the integration of science into the national security state, rather than the establishment of ideological tests for the content of science. Not even the "Oppenheimer affair" could halt or even slow the flow of scientists into military-related research.

■ China

As with the Soviets, science occupied a special position in the ideology of the Chinese Communist Party. The founders of the party in the 1920s turned to Marxism as a "scientific" explanation of history and communism as a natural course of social changes. Mao Zedong later developed an essentially instrumentalist ideology of science that might be called "revolutionary utilitarianism." "Revolutionary" referred to the goal of the Communist Party's science policy to ensure the political loyalty of the scientists, while "utilitarian" spoke to the pressure on scientists to produce immediate, practical results.

Even before their take-over of mainland China in 1949, Mao and his followers had launched a notorious "rectification" campaign in their stronghold in Yanan during the early 1940s. It resulted in a complete re-structuring of the scientific establishment under communist control. A number of scientists and science administrators had insisted on the priority of basic science in education and research. Now they were removed from their positions and punished in favor of those who advocated the re-orientation toward meeting immediate needs in production and military technology. Some of those suspected of harboring bourgeois thoughts—that is, thought not sufficiently revolutionary—were "sent down" to the countryside to learn from the peasants and thereby set an ominous precedent. As the historian James Reardon-Anderson points out, the narrow enforcement of revolutionary utilitarianism marginalized fundamental science "with long-term repercussions on the modernization of China."[80]

Ideological purification of the scientists, often Western-trained, started almost immediately in the People's Republic of China (PRC), and intensified during the "thought reform" campaign at the height of the Korean War in 1952–3. In a scene paralleling that in the U.S.A., war hysteria turned into a hunt for internal enemies, resulting in the suicides of a number of scientists in Shanghai.[81] Many scientists were accused of following bourgeois scientific theories, such as Mendelian genetics, cybernetics, resonance chemical theory, and Gestalt psychology, and forced to renounce them. The Anti-Rightist Campaign of 1957, however, outdid all these previous purges. Attacking those scientists who had criticized the party's mishandling and distrust of scientists as part of a conspiracy to overthrow the new government, Mao ordered the purge of hundreds of thousands of intellectuals, including scientists.[82] Many of the brightest scientists were thus taken away from science and education and placed in forced labor for many years. Despite periods of relaxation, the pressure on scientists for ideological purification never completely relented over the next two decades.[83]

Mao's distrust of scientists reached a crescendo during the Cultural Revolution of 1966–76. He and his supporters unleashed a harsh reign of terror by the radical Red Guards against anyone, including scientists, who could be accused of deviating from Mao's correct political line. Along with other intellectuals, scientists were again purged for their bourgeois ideology and their elitism; they had to be cleansed and reformed. Red Guards and other rebels took over scientific and educational institutions and stopped virtually all research. Scientists, especially those formerly in administrative positions, were criticized and persecuted, and sometimes beaten, tortured, and killed. By 1969, many scientists who survived the ordeal were sent to the countryside or factories to perform physical labor and help make a "people's science." Only after Mao's death in 1976 was it possible for a full-scale restoration of utilitarian science policy under the leadership of Deng Xiaoping.[84]

Scientists were usually purged in the Mao era not so much for the ideological content of their scientific theories as for their political opinions and even personal background, such as training in the West and working under the Nationalists before 1949. Yet, in a few cases, notably Lysenkoism, persecution did fall on those with the "wrong" beliefs. There were numerous cases where the pursuit of basic research by itself could bring on the indictment that one was ignoring the practical duties of a scientist and thus deviating from the correct Maoist model of integrating theory with practice. Perhaps the most striking feature of the treatment of Chinese scientists under Mao was the wide swings between liberalization and harsh tightening, which reflected the divisions within the Communist party leadership over the future course for China and the complexities of modernization.

In 1948–9, communist leaders encouraged scientists who had worked under the rival Nationalists to stay where they were, instead of following the fleeing Nationalist forces to Taiwan. After the establishment of the PRC in October 1949, many of these carry-over scientists attained important administrative positions in the reconstructed scientific institutions of the new regime, including the Chinese Academy of Sciences. Characteristically, however, the most important policies on personnel and research directions were determined by party officials and the few scientists who were also party members.[85] In the early 1950s, the party and government pursued a policy of encouraging scientific research and education, in part to persuade those who had worked under the Nationalists to stay with the new regime, and in part to attract those Chinese students and scientists training or working in the West to return to their homeland. Despite the various political purges in this period, by and large, the latter succeeded as thousands of them overcame obstacles in the West to return to China.[86]

Distrusting these carry-over and returnee scientists, however, the party launched efforts to train its own "red" experts almost immediately following the establishment of the PRC. Typically, the new recruits undertook narrowly focused undergraduate studies in one of the Chinese universities, which were radically restructured in the early 1950s according to the Soviet model to emphasize specialized technical fields, such as metallurgy or geology. The best of these students—in terms of both technical competence and political loyalty—were sent to the Soviet Union for graduate study. Upon their return, they were expected to become leaders in the Chinese scientific enterprise. For example, Zhou Guangzhou, a physics student, followed this path. He would later become a leader in the Chinese nuclear weapons project and president of the Chinese Academy of Sciences in the 1980s.[87]

During the early stage of the Cultural Revolution, there was little effort to train new scientists, except for what existed in the nuclear and military space projects. Most of the universities were shut down from 1966 to about 1971, with no students admitted or graduated. In 1971, universities were re-opened and operated under a radical new direction: freshmen were to come not from high school graduates based on national entrance examinations, as before, but from peasants, workers, and soldiers with practical experience but with junior high school preparation, selected on political criteria. The standards in this new educational regime proved so low that years later, following the end of the Cultural Revolution, the so-called peasant-worker-soldier students had to be retrained after graduation to reach university level.[88]

The experiment in recruitment and enlistment in science turned out to be a complete failure: it may have produced a "red," but not by any means "expert," generation. When the pragmatist party leader Deng Xiaoping, who was purged by Mao during the Cultural Revolution, returned to power following Mao's death in 1976, he brought the returnee/carryover generation of scientists back into power. The older scientists often bypassed the peasant-worker-soldiers of the Maoist era and began to train a new generation of scientists who came through a restored educational system. Many of the latter also began to pursue studies abroad, especially in the U.S.A. and Western Europe.

Deng's advocacy of utilitarianism continued to dominate Chinese science policy in the 1980s and 1990s as market-driven economic reform brought another wave of structural changes to Chinese science. A number of scientists did run afoul of the regime in this period, but, again, because of their political beliefs and activities, not their scientific theories. Fang Lizhi, the prominent astrophysicist and political dissident, lost his position as vice president of the University of Science and Technology of China, Hefei, and was expelled from the party when he was blamed for student unrest. However, in a move that echoed that of the Soviet Communist Party, Fang was allowed to work at an observatory in isolation from students. In fact, the regime intentionally publicized his research to indicate that it continued to value science although it discouraged political dissent.[89]

Despite the claims by the Red Guards and other Maoists at the time, the Cultural Revolution produced few, if any, ideologically correct scientific theories. Much of the energy of the radical Maoist theorists was focused on attacking what they viewed as bourgeois scientific beliefs within and outside of China, rather than constructing new ones. These included Albert Einstein's relativity theory, which the Maoists denounced as politically capitalistic and philosophically idealistic.[90] Perhaps the only plausible case of ideologically-correct-science in China took place on the eve of the Cultural Revolution.

On August 18, 1964, Mao Zedong invited several Chinese philosophers of science to his residence in Beijing's Forbidden City to chat about the philosophical implications of new theories of elementary particles. "The world is infinite," he asserted, "time and space are infinite. Space is infinite at both the macro and micro levels. Matter is infinitely divisible." Mao's comments, which were prompted by a recent article on elementary particles written by the Japanese Marxist and physicist Shoichi Sakata, led to widespread, officially sponsored discussions on philosophy of science. A number of physicists participated in the discussion, at the end of which they concluded that, based on Mao's version of dialectical materialism, elementary particles were not "elementary," but could be further divided into constituent parts—stratons—to signify the infinite stratification of matter.[91] Starting in 1965, these physicists, many of whom had participated in the making of the Chinese hydrogen bomb, began to construct a theoretical model of how stratons made up hadrons (protons and neutrons).[92]

At the time, there were already models of hadrons in the West, which assumed that they were composed of quarks, but it was not clear whether quarks were merely mathematical devices or real particles. The Beijing group claimed that its model differed from the quark model in that it was relativistic and assumed that there were real sub-particles that were made up of hadrons. However, the model was essentially compatible with, and later subsumed under the quark model, which eventually did assume that quarks represented not merely mathematical constructs but real physical entities in what is now called the Standard Model.[93]

There seems little doubt that the initiation and the philosophical interpretation of the straton theory had much to do with Mao's pronouncements on the infinite divisibility of matter in the Chinese context. Maoism provided both the political and ideological justification of a fairly esoteric branch of science that could easily have been branded a bourgeois exercise in an ivory tower. One might argue that these scientists would probably have received protection due to their contributions to the Chinese bomb project, as in the case of Soviet nuclear physicists described by David Holloway. But the fact that some of these scientists were criticized and not allowed to present the theory to an international gathering of scientists in Beijing after the start of the Cultural Revolution, despite the apparent ideological correctness of their work, seems to indicate that the "nuclear umbrella" worked less in Maoist China than in Stalinist Russia.[94] The straton model was thus more a case of protective coloration than a genuine scientific creation of dialectical materialism. Yet, perhaps more than any other scientific endeavors in China in the Maoist years, the straton model represented an attempt at ideologically-correct-science.

Not surprisingly, in comparison with the often critical reception of theoretical research by Maoist "revolutionary utilitarianism," applied research fared much better. In addition to the nuclear weapons project, which was both politically correct and utilitarian to the extreme, a number of other fields also received ample moral and material support in the Mao era. The pivotal Twelve-Year Plan in Science and Technology, formulated by hundreds of scientists under the leadership of Zhou Enlai in 1956, for example, identified about fifty-six applied research areas for heavy investment in resources and personnel. Only as an afterthought was "major theoretical problems in natural sciences" tagged on as the last item in the plan. Even in these basic research fields, applications were the major motivation.[95]

The Communist Party provided lavish rewards for achievements in applied research. Since the early 1990s, the faith in science and technology as the basis of China's modernization has flourished in China under Party leader Jiang Zemin, who had trained as an engineer. By applying modern science and technology to the economy and management, Jiang and his supporters hoped for both a robust economy and a stable social order under Communist rule.

Perhaps no modern state has purged and persecuted its scientists as ruthlessly and repeatedly as the People's Republic of China, but these waves of purges, as well as the intervals of relative tranquility which followed them, had little to do with ideologically-correct-science. Mao and his successors have wanted first and foremost obedience from their scientists. Ideologically correct conduct is part of this, but in this regard scientists have been no different than other Chinese.

■ Conclusion

If one assumes that the state (or forces or individuals within it) was imposing ideology on science and scientists, then this implies that science and ideology are separable. In fact, as the above examples demonstrate, even in the case of ICS science is not being determined by ideology, but also is not free of it. No political regime has ever tried consistently and comprehensively to impose ICS on its scientists. There have been individuals or portions of a regime that have tried to impose some ideologies on some aspects of science and on some scientists. However, such individual cases are often "overdetermined": there are alternative explanations other than ICS, both for the specific ideological attacks and the efficacy of the victims' resistance. On the other hand, ICS did happen. Just because the entire state was not behind it for the entire period does not contradict that. Although ICS in the ideal sense always failed, it also always had a significant effect, whether an intended one, or not.

No single ideology, including liberal democracy, has historically proven more effective than another in driving science or leading to intended results. Communist regimes appear to have been more likely to try and impose ideological standards on their science, perhaps because Marxism is so comprehensive a political philosophy. In other regimes, ideological attacks on science remained more crude and overt, like National Socialist calls to eliminate "Jewish Science" or anti-communists in the United States during the 1950s denouncing "internationalism" in science.

Although communist regimes were sometimes the most ruthless oppressors of their own scientists, they were also sometimes more flexible and pragmatic.[96] Thus in the Soviet Union and the Peoples' Republic of China, scientific representatives of the ideological enemy, the "bourgeois specialists," were kept on for a while, sometimes even pampered, in order to bridge the gap of time between the beginning of the revolution and the point at which the new regime had trained its own, ideologically acceptable (or more ideologically acceptable) scientists. In some cases, these holdovers from the previous regime held on long enough to shed their label and remain productive and sometimes integral parts of their national scientific efforts.

This would have been inconceivable in Hitler's Germany. The few exceptions of Jewish or part-Jewish scientists who managed to survive the Third Reich working for the regime are the exceptions that prove the rule. The overwhelming majority of "non-Aryan" scientists were purged because of their race and irregardless of their expertise or the scientific needs of the regime. Similarly, scientists suspected or accused of communist sympathies during the McCarthy period were damaged goods and not to be used by the government. It should also be noted that, whatever the actual loss of scientific manpower to Hitler's Germany and Eisenhower's America, in both cases there were sufficient numbers of ambitious, competent, and politically and ideologically acceptable scientists to fill the vacancies.

It is probably no coincidence that all of these examples also deal with military conflict or preparations for it. The French Revolution and National Socialism unleashed war, the Soviet Union and the People's Republic of China were born of war, Japan was fighting the Second World War, and in the 1950s America and the Soviet Union were battling in the Cold War. The militarization of society and of science bring the two closer together: the former's chances for success may hinge on the military potential of its science and technology; while the latter's service to its nation at war may entangle and immerse it more deeply in the regime's political and ideological goals. One pattern thus emerges quite often in the examples studied by this essay: the military potential of science and scientists outweigh and overrule attempts to purify science ideologically.

It was never easy to introduce ideologically driven and compliant science and technology. Purges of scientists or attacks on scientific theories on ideological grounds were often self-defeating and short-lived. Scientists sometimes suffered, but not because they were scientists, rather because they were part of a greater ideological or racial group perceived by the regime to be a potential threat. When challenged, scientists did not defend themselves by claiming the ideological neutrality of their work and demanding intellectual freedom. Instead they strove to blunt attacks by either winning over their critics or enlisting other patrons. Of course, this meant that they had to work closely, or at least more closely with at least some forces in the state and demonstrate their usefulness to them.

The end result was usually partial success for the established (and now sometimes embattled) scientific community, for their critics were eventually silenced and they did safeguard some of their professional prerogatives, but also partially entailed an accommodation to, and collaboration with ideological aspects of the regime, sometimes directly related to their scientific work, other times in a more general form.

This is not what the sociologist Robert Merton or physicist Samuel Goudsmit predicted in the aftermath of the Second World War and the beginning of the Cold War.[97] Science is not especially suited to democracy, as Goudsmit claimed, and is able to compromise some of Merton's norms for scientific work. Indeed it is striking, that, despite the great differences between these often very different examples, most of the scientists responded in a similar way when their apolitical science was threatened. ICS tells us that (1) no matter how ruthless, totalitarian, racist, or intolerant a regime might be, when it needs its scientists, it will do what it has to in order to harness them; and (2) whether they support the regime whole-heartedly or not, most scientists, or perhaps better put, scientific communities, will do what they have to in order to be able to do science. Thus science is independent of particular political and ideological regimes: just not in the way most people believe it is.

Notes

1. Society of Fellows, Harvard University and History Department, Princeton University; Department of History, Bowling Green State University; Union College; California State Polytechnic University, Pomona. We would like to thank the colleagues who read and commented on a preliminary version of our manuscript: Mitchell Ash, Charles C. Gillispie, Loren Graham, Alexei Kojevnikov, and Jessica Wang.
2. Jessica Wang, *American Science in an Age of Anxiety: Scientists, Anti-Communism, and the Cold War* (Chapel Hill, NC: University of North Carolina Press, 1998).
3. Doug Weiner, *Models of Nature* (Bloomington, IN: Indiana University Press, 1988).
4. Harold T. Parker, "French Administrators and French Scientists during the Old Regime and the Early Years of the Revolution," in Richard Herr and Harold T. Parker (eds), *Ideas in History* (Durham, NC: Duke University Press, 1965), 85–109.
5. Roger Hahn, *The Anatomy of a Scientific Institution: The Paris Academy of Sciences, 1666–1803* (Berkeley, CA: University of California Press, 1971); Charles C. Gillispie, *Science and Polity in France at the End of the Old Regime* (Princeton, NJ: Princeton University Press, 1980), 81–99.
6. Keith Michael Baker, *Condorcet* (Chicago: University of Chicago Press, 1975), 350–2.
7. Henry Moss, "Scientists and Sans-culottes: The Spread of Scientific Literacy in the Revolutionary Year II," *Fundamenta Scientiae*, 4 (1983): 101–15; Gillispie, *Science and Polity in France at the End of the Old Regime,* 143–84.
8. Ken Alder, *Engineering the Revolution: Arms and Enlightenment in France, 1763–1815* (Princeton, NJ: Princeton University Press, 1997), ch. 8; Terry Shinn, *L'École Polytechnique, 1794–1914* (Paris: Presses de la fondation nationale des sciences politiques, 1980); Ambroise Fourcy, *Histoire de l'Ecole*

Polytechnique (Paris: Belin, 1987); Nicole et Jean Dhombres, *Naissance d'un nouveau pouvoir: Sciences et Savants en France, 1793–1824* (Paris: Payot, 1989); Janis Langins, "The *Ecole Polytechnique* and the French Revolution: Merit, Militarization, and Mathematics," *LLULL*, 13 (1990): 91–105; and Janis Langins, *La République avait besoin de savants* (Paris: Belin, 1987).

9. For a short account, see Langins, "The *Ecole Polytechnique*". More detailed treatments are provided by the Alder and Langins works cited above.

10. See Charles C. Gillispie, "The *Encyclopédie* and the Jacobin Philosophy of Science: A Study in Ideas and Consequences," in Marshall Clagett (ed.), *Critical Problems in the History of Science* (Madison, WI: University of Wisconsin Press, 1959), 255–89; and L. Pearce Williams, "The Politics of Science in the French Revolution," in Clagett, 291–320.

11. Jessica Riskin, "Rival Idioms for a Revolutionized Science and a Republican Citizenry," *Isis*, 89 (1998): 203–32.

12. Charles C. Gillispie, "De l'Histoire naturelle à la biologie: Relations entre les programmes des recherche de Cuvier, Lamarck, et Geoffrey Saint-Hilaire," in *Collecter, observer, classer* (forthcoming).

13. J.W. Dauben, "Marat: His Science and the French Revolution," *Archives Internationales de l'Histoire des Sciences*, 22 (1969): 235–61; Gillispie, *Science and Polity in France at the End of the Old Regime*, 290–330.

14. See Hahn, *passim*, for references to primary literature on the Jardin.

15. Loren R. Graham, *The Soviet Academy of Sciences and the Communist Party, 1927–1932* (Princeton, NJ: Princeton University Press, 1967); and Alexander Vucinich, *Empire of Knowledge: The Academy of Sciences of the USSR (1917–1970)* (Berkeley, CA: University of California Press, 1984).

16. Graham, *The Soviet Academy of Sciences*, 24–79; and Kendall E. Bailes, *Technology and Society under Lenin and Stalin: Origins of the Soviet Technical Intelligentsia, 1917–1941* (Princeton, NJ: Princeton University Press, 1978).

17. Lynn Mally, *Culture of the Future: The Proletkult Movement in Revolutionary Russia* (Berkeley, CA: University of California Press, 1990). Katerina Clark, "The Changing Image of Science and Technology in Soviet Literature," in Loren R. Graham (ed.), *Science and the Soviet Social Order* (Cambridge, MA: Harvard University Press, 1990), 266–67.

18. Mally, chapter 6.

19. On specialists during this period, see Bailes.

20. Jeremy R. Azrael, *Managerial Power and Soviet Politics* (Cambridge, MA: Harvard University Press, 1966). On the economics and politics of the NEP period, see Stephen F. Cohen, *Bukharin and the Bolshevik Revolution: A Political Biography, 1888–1938* (Oxford: Oxford University Press, 1980), and Robert V. Daniels, *Conscience of the Revolution: Communist Opposition in Soviet Russia* (Boulder, CO: Westview Press, 1988).

21. On the tense compromise between ideological promotion of young Soviet acolytes like Landau and the surviving tsarist professorate, see Karl Hall, "Purely Practical Revolutionaries," (Ph.D., Harvard University, 1999).

22. For a detailed account of the Shakhty Trial and the subsequent Industrial Party Affair, see Bailes, chapter 3.

23. David Holloway, *Stalin and the Bomb* (New Haven: Yale Universiry Press, 1994).

24. On the education system generally in this period, see Sheila Fitzpatrick, *The Commissariat of Enlightenment: Soviet Organization of Education and the Arts under Lunacharsky, October 1917–1921* (Cambridge: Cambridge University Press, 1970). On the reform of institutes, see Loren R. Graham, "The Formation of Soviet Research Institutes: A Combination of Revolutionary Innovation and International Borrowing," *Social Studies of Science, 5* (1975): 303–29. On Leningrad Physico-Technical Institute in this period, see Paul Josephson, *Physics and Politics in Revolutionary Russia* (Berkeley, CA: University of California Press, 1991).

25. See Graham, *The Soviet Academy of Sciences,* 24–79; Vucinich, 72–129; and Azrael, chapters 3–4.

26. Bailes.

27. Sheila Fitzpatrick, *The Cultural Front: Power and Culture in Revolutionary Russia* (Ithaca, NY: Cornell University Press, 1992).

28. David Joravsky, *Soviet Marxism and Natural Science, 1917–1932* (London: Routledge and Kegan Paul, 1961).

29. Loren R. Graham, *Science in Russia and the Soviet Union: A Short History* (Cambridge: Cambridge University Press, 1993), ch. 5.

30. Loren R. Graham, *Science, Philosophy, and Human Behavior in the Soviet Union* (New York: Columbia University Press, 1987), chaps 1–2.

31. David Joravsky, *The Lysenko Affair* (Cambridge, MA: Harvard University Press, 1970); Krementsov; Valery N. Soyfer, *Lysenko and the Tragedy of Soviet Science* (New Brunswick, NJ: Rutgers University Press, 1994); Zhores A. Medvedev, *The Rise and Fall of T.D. Lysenko,* (New York: Columbia University Press, 1969); David Joravsky, "Soviet Marxism and Biology Before Lysenko," *Journal of the History of Ideas,* 20 (1959): 85–104; Dominique Lecourt, *Proletarian Science? The Case of Lysenko* (Norfolk, VA: NLB, 1977); Nikolai L. Krementsov, *Stalinist Science* (Princeton, NJ: Princeton University Press, 1997).

32. Kirill O. Rossianov, "Stalin as Lysenko's Editor: Reshaping Political Discourse in Soviet Science," *Configurations,* 3 (1993): 439–56; and Kirill O. Rossianov, "Editing Nature: Joseph Stalin and the 'New' Soviet Biology," *Isis,* 84 (1993): 728–45.

33. Alexei Kojevnikov, "Rituals of Stalinist Culture at Work: Science and the Games of Intraparty Democracy circa 1948," *Russian Review,* 57, 1 (1998): 25–52.

34. Mitchell Ash and Alfons Söllner, "Forced Migration and Scientific Change after 1933," in Mitchell Ash and Alfons Söllner (eds), *Forced Migration and Scientific Change: Emigré German-Speaking Scientists and Scholars after 1933* (Cambridge: Cambridge University Press, 1996), 1–19.

35. Alan Beyerchen, *Scientists under Hitler: Politics and the Physics Community in the Third Reich* (New Haven, CT: Yale University Press, 1977).

36. Rudolf Vierhaus and Bernhard vom Brocke (eds), *Forschung im Spannungsfeld von Politik und Gesellschaft–Geschichte und Struktur der Kaiser-Wilhelm/Max-Planck-Gesellschaft* (Stuttgart: DVA, 1990); Kristie Macrakis, *Surviving the Swastika: Scientific Research in Nazi Germany* (Cambridge, MA: Harvard University Press, 1993).

37. Mark Walker, *Nazi Science* (New York: Plenum, 1995).

38. Robert Proctor, *Racial Hygiene: Medicine under the Nazis* (Cambridge, MA: Harvard University Press, 1988); Paul Weindling, *Health, Race, and German Politics* between *National Unification and Nazism, 1870–1945* (Cambridge: Cambridge University Press, 1989); Uwe Hoßfeld, "Staatsbiologie, Rassenkunde und Moderne Sythese in Deutschland während der NS-Zeit," in Rainer Brömer, Uwe Hoßfeld, and Nicolaas Rupke (eds), *Evolutionsbiologie von Darwin bis heute* (Berlin: VWB, 2000), 249–305; Uwe Hoßfeld and Thomas Junker, "Synthetische Theorie und 'Deutsche Biologie': Einführender Essay," in Brömer, Hoßfeld, and Rupke, 231–48; Thomas Junker, "Synthetische Theorie, Eugenik und NS-Biologie," in Brömer, Hoßfeld, and Rupke, 307–60.

39. Walker, *Nazi Science.*

40. Jacques Pauwels, *Women, Nazis, and Universities: Female University Students in the Third Reich* (New Haven, CT: Greenwood Press, 184).

41. Alexander Mitscherlich and Fred Mielke, *Doctors of Infamy: The Story of the Nazi Medical Crimes* (New York: Henry Schuman, 1949); Robert J. Lifton, *The Nazi Doctors: Medical Killing and the Psychology of Genocide* (New York: Basic Books, 1986); Benno Müller-Hill, *Murderous Science: Elimination by Scientific Selection of Jews, Gypsies, and Others, Germany 1933–1945* (Oxford: Oxford University Press, 1985).

42. For biology, Ute Deichmann, *Biologists under Hitler* (Cambridge, MA: Harvard University Press, 1996); for physics, Beyerchen, David Cassidy, *Uncertainty: The Life and Science of Werner Heisenberg* (New York: Freeman, 1991), Mark Walker, *German National Socialism and the Quest for Nuclear Power, 1939–1949* (Cambridge: Cambridge University Press, 1989), and Walker, *Nazi Science;* for mathematics, Herbert Mehrtens, "Ludwig Bieberbach and 'Deutsche Mathematik,'" in E. Phillips (ed.), *Studies in the History of Mathematics* (Washington, DC: American Mathematical Association, 1987), 195–247; for psychology see Ulfried Geuter, *The Professionalization of Psychology in Nazi Germany* (Cambridge: Cambridge University Press, 1992), Mitchell Ash, "From 'Positive Eugenics' to Behavioral Genetics: Psychological Twin Research under Nazism and Since," *Historia Pedagogica—International Journal*

of the History of Education, Supplementary Series, 3 (1998): 335–58, and Mitchell Ash, "Constructing Continuities: Kurt Gottschaldt und Psychological Research in Nazi and Socialist Germany," in Kristie Macrakis and Dieter Hoffmann (eds), *Science and Socialism in the G.D.R. in Comparative Perspective* (Cambridge, MA.: Harvard University Press, 1999), 286–301, 360–65; for chemistry Ute Deichmann, *Flüchetn, Mitmachen, Vergessen* (Weinheim: Wiley-VCH, 2001); for technology, Karl' Heinz Ludwig, *Technik und Ingenieure im Dritten Reich* (Düsseldorf: Droste Verlag, 1974) and Helmut Maier, "National sozialistische Technikideologie und die Politisierung des 'Technikerstandes': Fritz Todt und die Zeitschrift 'Deutsche Technik,'" in Burkhard Dietz, Michael Fessner, and Helmut Maier (eds), *Technische Intelligent und "Kulturfaktor Technik"* (Münster: Waxmann, 1996), 253–68.

43. Cassidy; Walker, *Nazi Science.*

44. Michael Neufeld, *The Rocket and the Reich: Peenemünde and the Coming of the Ballistic Missile Era* (New York: Free Press, 1995).

45. Helmuth Trischler, *Luft- und Raumfahrtforschung in Deutschland 1900–1970* (Frankfurt am Main: Campus Verlag, 1992); Ulrich Albrecht, "Military Technology and National Socialist Ideology," in Renneberg and Walker (eds), 88–125, 358–63.

46. Walker, *German National Socialism and the Quest for Nuclear Power.*

47. For physicians, Lifton, Weindling, Proctor, for biologists, Deichmann, Junker; for psychologists, Geuter, Mitchell Ash, "Denazifying Scientists and Science," in Matthias Judt and Burghard Ciesla (eds), *Technology Transfer out of Germany* (Amsterdam: Harwood, 1996), 61–80; for anthropologists, Hossfeld; for geographers, Mechtild Rössler, "'Area Research' and 'Spatial Planning' from the Weimar Republic to the German Federal Republic: Creating a Society with a Spatial Order under National Socialism," in Renneberg and Walker (eds), 126–38, 363–6.

48. Two examples are Dr. Pak Ch'ul Jai, an X-ray physicist, and Dr. Lee Tai Kyu, a professor of chemistry, both of whom held positions at Kyoto Imperial University during the war. See Walter E. Grunden, "Hungnam and the Japanese Atomic Bomb: Recent Historiography of a Postwar Myth," *Intelligence and National Security,* 13 (Summer 1998): 32–60.

49. Yukawa was awarded the Nobel Prize in physics in 1949. Taketani Mitsuo, "Methodological Approaches in the Development of the Meson Theory of Yukawa in Japan," in Nakayama Shigeru, David L. Swain, and Yagi Eri (eds), *Science and Society in Modern Japan: Selected Historical Sources* (Tokyo: University of Tokyo Press, 1973), 24–38.

50. Hiroshige Tetu, "The Role of the Government in the Development of Science," *Journal of World History,* 9 (1965): 320–39. Itakura Kiyonobu and Yagi Eri, "The Japanese Research System and the Establishment of the Institute of Physical and Chemical Research," in Nakayama, Swain, and Yagi (eds), *Science and Society in Modern Japan,* 158–201. Kamatani Chikayoshi, "The History of Research Organization in Japan, *Japanese Studies in the History of Science,* 2 (1963): 1–79.

51. S. Watanabe, "How Japan Has Lost a Scientific War" (September 1945), File: "Historical Information," RG 38, Box 111, U.S. National Archives, Washington, DC.

52. Yomiuri Shimbunsha (ed.), "Nihon no genbaku" (Japan's Atomic Bomb), in *Shôwa shi no Tennô* (The Emperor in Shôwa History), vol. 4 (Tokyo: Yomiuri Shimbunsha, 1968), 78–229.

53. John W. Dower, "'NI' and 'F': Japan's Wartime Atomic Bomb Research," *Japan in War and Peace: Selected Essays* (New York: New Press, 1993), 55–100. Morris Fraser Low, "Japan's Secret War? 'Instant' Scientific Manpower and Japan's World War II Atomic Bomb Project," *Annals of Science,* 47 (1990): 347–60.

54. Walter E. Grunden, "Science under the Rising Sun: Weapons Development and the Organization of Scientific Research in World War II Japan" (Ph.D., University of California at Santa Barbara, 1998).

55. Tessa Morris-Suzuki, *The Technological Transformation of Japan: From the Seventeenth to the Twenty-First Century* (Cambridge: Cambridge University Press, 1994), 144–5.

56. Morris-Suzuki, 145.

57. Excerpts of Lieutenant-General Tada Reikichi, President of the Board of Technology, from Captain George B. Davis to Major Francis J. Smith, "Transmittal of Items from the 'Daily Digest of World Broadcasts'" (27 July 1945), "Japan, Misc" RG 77, Entry 22, Box 173, Folder #44.70, U.S. National Archives, College Park, MD; See also, Major H.K. Calvert to Major F.J. Smith, "Japanese Militarists

Want Miracle Weapon" (29 May 1945) "Japan, Misc" RG 77, Entry 22, Box 173, Folder #44.70, U.S. National Archives, College Park, MD.

58. Peter Williams and David Wallace, *Unit 731: Japan's Secret Biological Warfare in World War II* (New York: The Free Press, 1989).

59. For a concise treatment of McCarthyism in general, see Ellen Schrecker, *No Ivory Tower: McCarthyism and the Universities* (New York: Oxford University Press, 1986) and Ellen Schrecker, *The Age of McCarthyism: A Brief History with Documents* (Boston: Bedford Books, 1994).

60. Lawrence Badash, *Scientists and the Development of Nuclear Weapons: From Fission to the Limited Test Ban Treaty, 1939–1963* (Atlantic Highlands, NJ: Humanities Press, 1995), 102.

61. Paul Forman, "Behind Quantum Electronics: National Security as Basis for Physical Research in the United States, 1940–1960," *Historical Studies in the Physical and Biological Sciences,* 18 (1987): 149–229.

62. Richard G. Hewlett and Jack M. Holl, *Atoms for Peace and War, 1953–1961: Eisenhower and the Atomic Energy Commission* (Berkeley, CA: University of California Press, 1989), ch. 4, "The Oppenheimer Case," 73–112.

63. Wang, *Anti-Communism*, 289–95.

64. Wang, *Anti-Communism*, 274–9.

65. "Scientists Affirm Faith in Oppenheimer," *Bulletin of the Atomic Scientists,* 10 (May 1954): 188–91. "Scientists Express Confidence in Oppenheimer," *Bulletin of the Atomic Scientists,* 10 (September 1954): 283–6.

66. Steven E. Ambrose, *Eisenhower: The President* (New York: Simon and Schuster, 1984), 166.

67. Eisenhower to Robert E. Sherwood, April 21, 1954, in *The Diaries of Dwight D. Eisenhower,* reel 4, 193. See also Dwight D. Eisenhower, *Mandate for Change, 1953–1956* (Garden City, NY: Doubleday and Company, Inc., 1963), 314.

68. Hewlett and Holl, 111–12.

69. Wang, *Anti-Communism*.

70. Interview with Herbert York by Zuoyue Wang, La Jolla, CA, July 18, 1992.

71. Daniel Kevles, "Cold War and Hot Physics: Science, Security, and the American State, 1945–56," *Historical Studies in the Physical and Biological Sciences,* 20 (1990): 239–64.

72. Wang, *Anti-Communism*, 261–2, 281.

73. Schrecker, *No Ivory Tower,* and Schrecker, *The Age of McCarthyism;* interview with Wolfgang Panofsky by Zuoyue Wang, Stanford, CA, March 5, 1992.

74. Schrecker, *No Ivory Tower;* Wang, *Anti-Communism*, 254–261.

75. William Lanouette, *Genius in the Shadows: A Biography of Leo Szilard, the Man Behind the Bomb* (Chicago: University of Chicago Press, 1994).

76. Wang, *Anti-Communism*, 277.

77. Wang, *Anti-Communism*, 289–295, Schrecker, *The Age of McCarthyism,* 92–94.

78. Zuoyue Wang, *In the Shadow of Sputnik: American Scientists and Cold War Science Policy* (forthcoming). Wang, *Anti-Communism*.

79. Schrecker, *No Ivory Tower;* Schrecker, *The Age of McCarthyism,* 92.

80. James Reardon-Anderson, *The Study of Change: Chemistry in China, 1840–1949* (Cambridge: Cambridge University Press, 1991), 339–64, on 359.

81. Yao Slurping *et al., Zhongguo kexueyuan* (Chinese Academy of Sciences), 3 volumes (Beijing: Contemporary China Press, 1994), vol. 1, 26–7.

82. Jonathan Spence, *The Search for Modern China* (New York: W.W. Norton 1990) 572.

83. Yao *et al.*, 87–9.

84. See, for example, Peter Neushul and Zuoyue Wang, "Between the Devil and the Deep Sea: C.K. Tseng, Ocean Farming, and the Politics of Science in Modern China," *Isis*, 91, 1 (2000): 59–88.

85. See, for example, several articles in the Chinese official journal *Bai nian chao* (Hundred Year Tide), June 1999, especially Li Zhenzhen, "Interview with Yu Guangyuan and Li Peishan," 23–30, and "Interview with Gong Yuzhi," 31–27.

86. See Jin Chongji, *Zhou Enlai zhuan* (A Biography of Zhou Enlai), 2 volumes (Beijing: Central Documentation Press, 1998), vol. 1, 234.

87. Dai Minghua, *et al*, "Zhou Guangzhao" in Lu Jiaxi (ed. in chief), *Zhongguo xiandai kexuejia zhuanji* (Biographies of Modern Chinese Scientists), 6 volumes (Beijing: Science Press, 1991–4), vol. 6, 187–96.

88. Wu Heng, *Keji zhanxian wushinian* (Fifty Years on the Scientific and Technological Front) (Beijing: Science and Technology Documentation Press, 1992), 348–9.

89. See H. Lyman Miller, *Science and Dissent in Post-Mao China: The Politics of Knowledge* (Seattle: University of Washington Press, 1996).

90. Xu Liangying and Qiu Jingcheng, "Guanyu wuoguo 'wenhua dageming' shiqi pipan ai-in-si-tan he xiangduilun de chubu kaocha" (A Preliminary Study of the Movement to Denounce Einstein and Relativity During the Period of the Great Cultural Revolution in Our Country), in Xu (ed. in chief), *Ai-in-si-tan yanjiu* (Einstein Studies), 1 (1989): 212–50.

91. Gong Yuzhi, "Mao Zedong yu zhiran kexue" (Mao Zedong and the Natural Sciences), in Gong, *Zhiran bianzhengfa zai zhongguo* (Dialectics of Nature in China) (Beijing: Peking University Press, 1996), 87–112.

92. He Zuoxiu, "A History of the Establishment of the Straton Model," *Yuanshi zhiliao yu yanjiu* (Documentation and Research in the History of the Chinese Academy of Sciences), 1994, issue no. 6, 16–30.

93. He.

94. Holloway.

95. See "1956–1967 nian kexue jishu fazhan guihua gangyao" (Outline of a Long-Term Plan for the Development of Science and Technology, 1956–1967), in Chinese Communist Party Central Documentation Institute (ed.), *Jianguo yilai zhongyao wenxián xuanbian* (Selected Key Documents Since the Founding of the Country) (Beijing: Central Documentation Press, 1992–), vol. 9, 436–535.

96. Also see for the example of the German Democratic Republic the following articles: Mitchell Ash, "Wissenschaft, Politik und Modernität in der DDR: Ansätze zu einer Neubetrachtung," in K. Weisemann, P. Kroener, and R. Toellner (eds), *Wissenschaft und Politik: Genetik und Humangenetik in der DDR (1949–1989)* (Münster: LIT-Verlag, 1997), 1–26; Mitchell Ash, "1933, 1945 und 1990—drei Bruchstellen in der Geschichte der deutschen Universität," in A. Söllner (ed.) *Ostblicke—Perspektiven der Hochschulen in den neuen Bundesländem* (Opladen: Westdeutscher Verlag, 1998), 212–37; Mitchell Ash, "Scientific Changes in Germany 1933, 1945 and 1990: Towards a Comparison," *Minerva*, 37 (1999): 329–54.

97. See the introduction to this volume, as well as Jessica Wang, "Merton's Shadow: Perspectives on Science and Democracy since 1940," *Historical Studies in the Physical and Biological Sciences,* 30, No. 1 (1999): 279–306, and David Hollinger, "The Defense of Democracy and Robert K. Merton's Formulation of the Scientific Ethos," in David Hollinger, *Science, Jews, and Secular Culture: Studies in Mid-Twentieth Century American Intellectual History* (Princeton, NJ: Princeton University Press, 1996), ch. 5.

WWII Fighter Jet
Image © Peter Elvidge, 2010. Under license from
Shutterstock, Inc.

British Tank from WWI
Image © Arthur Tiutenko, 2010. Under license
from Shutterstock, Inc.

WWI Machine Gun
Image © Alistair Cotton, 2010. Under license from
Shutterstock, Inc.

Atomic explosion with shockwave
Image © Leigh Prather, 2010. Under license
from Shutterstock, Inc.

WWI Trench Warfare
Image © Andre Nantel, 2010. Under license from
Shutterstock, Inc.

Auschiwitz-Birkeneau camp, Poland.
Nazi Gas Chambers
Image © posztos, 2010. Under license from
Shutterstock, Inc.

Reproduction of German Folker Dr 1 Triplane
Image © Richard Thornton, 2010. Under license from
Shutterstock, Inc.

Soldier in WWI Gas Mask
Image © Losevsky Pavel, 2010. Under
license from Shutterstock, Inc.

WWII Radar Station
Image © carstheis, 2010. Under license
from Shutterstock, Inc.

Section 3

Communications Technologies

Introduction: How Users and Non-Users Matter

Reading 9

Nelly Oudshoorn and Trevor Pinch

New uses are always being found for familiar technologies. Sometimes these changes in use are dramatic and unexpected. Before September 11, 2001, no one foresaw that an airliner could be turned by a small number of its occupants into a giant Molotov cocktail. After the Gulf War of 1991, it was discovered that an effective way to put out oil-rig fires was to strap down captured Mig jet fighters and blow out the fires using their exhaust. Such examples remind us that we can never take the use of a technology for granted.

Susan Douglas (1987) has pointed out how amateur operators discovered new uses to which the emerging technology of radio could be put, and how commercial operators soon followed the amateurs' lead. Claud Fischer (1992) and Michele Martin (1991) have drawn attention to the use of the telephone by rural women to overcome their isolation—a use not foreseen by telephone companies, which conceived of the telephone mainly as a business instrument.

Our concern in this book is with the role of users in the development of technology in general. We are interested in how users consume, modify, domesticate, design, reconfigure, and resist technologies. In short, our interest is in whatever users do with technology.

There is no one correct use for a technology. "What is an alarm clock for?" we might ask. "To wake us up in the morning," we might answer. But just begin to list all the uses to which an alarm clock can be put and you see the problem. An alarm clock can be worn as a political statement by a rapper; it can be used to make a sound on a Pink Floyd recording; it can be used to evoke laughter, as Mr. Bean does in one of his comic sketches as he tries to drown his alarm clock in his bedside water pitcher; it can be used to trigger a bomb; and, yes, it can be used to wake us up. No doubt there are many more uses. Of course, there may be one dominant use of a technology, or a prescribed use, or a use that confirms the manufacturer's warranty, but there is no one essential use that can be deduced from the artifact itself. This is an axiomatic assumption for the scholars whose work we collect here. All the contributors follow the research path of studying technologies in their "context of use"—the society and the web of other artifacts within which technologies are always embedded. In short, we look at how technologies are actually used in practice.

In addition to studying what users do with technology, we are interested in what technologies do to users. Users of technologies do not arrive de novo. Think of the camera. When George Eastman developed his revolutionary new technology of roll film and a cheap camera, he had one outstanding problem: There were as yet no users for it. Photography was seen as a high-end activity practiced by a small group of skilled professionals. Eastman had to define explicitly who the new users might be, and he had to figure out how to recruit them to his new technology. He had to redefine photography and the camera. After he did, photography became something that anyone could participate in, and cameras became usable by all (Jenkins 1975). Working out who the new users are and how they will actually interact with a new technology is a problem familiar to many innovators of new technologies. Some fields, including information technology, are particularly cognizant of the problem of users. It has long been recognized that the most sophisticated and complex computer hardware and software will come to naught if users don't known how to use them. Studies of human-computer interaction, of work practices, and of user interfaces are often carried out by the computer industry, and they have become important not only for that industry but also for developing new ideas of how the user-technology nexus should be conceptualized (Suchman 1994; Woolgar 1991).

One important research question addressed in this book is how users are defined and by whom. For instance, are users to be conceived of as isolated autonomous consumers, or as self-conscious groups? How do designers think of users? Who speaks for them, and how? Are users an important new political group, or a new form of social movement? In short, what general lessons are to be drawn from a renewed focus on users in today's technologically mediated societies?

Different Approaches to Users

Users and technology are too often viewed as separate objects of research. This book looks for connections between the two spheres. Users and technology are seen as two sides of the same problem—as co-constructed. The aim is to present studies of the co-construction of users and technologies that go beyond technological determinist views of technology and essentialist views of users' identities.

In this introduction we discuss several influential approaches to user-technology relations, focusing in particular on the conceptual vocabulary developed within the different approaches and on the similarities and differences between them.

The SCOT Approach: Users as Agents of Technological Change

In the 1980s and the 1990s, the old view of users as passive consumers of technology was largely replaced in some areas of technology studies, and along with it the linear model of technological innovation and diffusion. One of the first approaches to draw attention to users was the social construction of technology (SCOT) approach.

Pinch and Bijker (1984), in defining the SCOT approach, conceived of users as a social group that played a part in the construction of a technology. Different social groups, they noted, could construct radically different meanings of a technology. This came to be known as a technology's interpretive flexibility. In a well-known study of the development of the bicycle, it was argued that elderly men and women gave a new meaning to the high-wheeled bicycle as the "unsafe" bicycle, and that this helped pave the way for the development of the safety bicycle. The SCOT approach specifies a number of closure mechanisms—social processes whereby interpretative flexibility is curtailed. Eventually, a technology stabilizes, interpretative flexibility vanishes, and a predominant meaning and a predominant use emerge (Bijker and Pinch 1987; Bijker 1995). The connection between designers and users was made more explicit with the notion of a technological frame (Bijker 1995). Users and designers could be said to share a technological frame associated with a particular technology.

Many of the classic SCOT studies were of the early stages of technologies. For example, there were studies of how the bicycle, fluorescent lighting, and Bakelite moved from interpretative flexibility to stability. Early on, social groups were seen as the shaping agents. Not until later, with notions such as that of sociotechnical ensembles, did SCOT fully embrace the idea of the co-construction or mutual shaping of social groups and technologies (Bijker 1995b). The SCOT approach was rightly criticized for its rather cavalier attitude toward users—it closed down the problem of users too early, and it did not show how users could actively modify stable technologies (Mackay and Gillespie 1992). Kline and Pinch (1996) remedied this with their study of how a stable technology, the Model T automobile, could be appropriated and redesigned by groups such as farmers who used cars as stationary power sources. Kline and Pinch referred so such users as "agents of technological change." Also attempting to correct SCOT's neglect of gender, Kline and Pinch argued that users should be studied as a crucial location where often-contradictory genders identities and power relationships were woven around technologies. Bijker (1995) argued for a semiotic conception of power whereby power is embedded and mediated by artifacts as well as by frames and social groups. However, this semiotic notion of power (like most semiotic approaches within technology studies) seems inevitably to leave out invisible actors and social groups, which in the SCOT approach might be termed "non-relevant social groups."

Feminist Approaches: Diversity and Power

Feminist scholars have played a leading role in drawing attention to users. Their interest in users reflects concerns about the potential problematic consequences of technologies for women and about the absence of women in historical accounts of technology. Since the mid 1980s, feminist historians have pointed to the neglect of women's role in the development of technology. Because women were historically underrepresented as innovators of technology, and because historians of technology often focused exclusively on the design and production of technologies, the history of technology came to be dominated by stories about men and their machines. Moreover, these stories represented a discourse in which gender was invisible. Historian did not consider it relevant in settings where women were absent, thus reinforcing the view that men had no gender. Feminist historians suggested that focusing on users and use rather than on engineers and design would enable historians to go beyond histories of men inventing and mastering technology (Wajcman 1991; Lerman et al. 1997). In response to this criticism, users were gradually included in the research agenda of historians of technology. This "turn to the users" can be traced back to Ruth Schwartz Cowan's exemplary research on user-technology relations. In the late 1970s, Cowan brought the fields of history of technology and women's history together, emphasizing that women as users of technology perceive technological change in significantly different ways from men (Pursell 2001). Cowan's notion of "the consumption junction," defined as "the place and time at which the consumer makes choices between competing technologies" (Cowan 1987: 263), was a landmark. Cowan argued that focusing on the consumer and on the network relations in which the consumer is embedded enables historians and sociologists of technology to improve their understanding of the unintended consequences of technologies in the hands of users. Focusing on users would enrich the history of technology with a better understanding of the successes and failures of technologies (ibid.: 279). In contrast to actor-network theory (which we will discuss below), Cowan urged historians and sociologists of technology to choose the user, rather than the artifact or the technologist, as a point of departure in network analyses of technology, and to look at networks from the consumer's point of view (ibid.: 262). The scholarship that Cowan inspired rejects the idea that science and technology begin or end with the actions of scientists and engineers. Scholars in the field of Science and Technology Studies (STS) were urged to follow technologies all the way to the users (Rapp 1998: 48). An exemplary study is Cynthia Cockburn and Susan Ormrod's 1993 book on the microwave oven in the United Kingdom, which analyzes the design, the production, and the marketing as well as the use of a new technology.

Gender studies, like technology studies in general, reflect a shift in the conceptualization of users from passive recipients to active participants. In the early feminist literature, women's relation to technology had been conceptualized predominantly in terms of victims of technology. The scholarship of the last two decades, however, has emphasized women's active role in the appropriation of technology. This shift in emphasis was explicitly articulated in the first feminist collection of historical research on technology, *Dynamos and Virgins Revisited* (Trescott 1979), which included a section on "women as active participants in technological change" (Lerman et al. 1997: 11). The authors of the essays in that section argued that feminists should go beyond representations of women as essentially passive with respect to technology. Having accepted that challenge, feminist historians, anthropologists, and sociologists have published numerous accounts of how women shape and negotiate meanings and practices in technology, including studies of the relationship between reproductive technologies and women's health and autonomy, of the gendered medicalization of bodies, of women's relations to computers and the impact of computer technologies on women's work, of the consequences of household technologies for women's lives, and of the exclusion of women from technologies. Granting agency to users, particularly women, can thus be considered central to the feminist approach to user-technology relations.

Another important concept in feminist studies of technology is diversity. As Cowan (1987) suggested, users come in many different shapes and sizes. Medical technologies, for example, have a wide variety of users, including patients, health professionals, hospital administrators, nurses, and patients' families. "Who is the user?" is far from a trivial question. The very act of identifying specific individuals or groups as users may facilitate or constrain the actual roles of specific groups of users in shaping the development and use of technologies. Different groups involved in the design of technologies may have different views of who the user might or should be, and these different groups may mobilize different resources to inscribe their views in the design of technical objects (Oudshoorn et al., forthcoming). And these different type of users don't necessarily imply homogeneous categories. Gender, age, socio-economic, and ethnic differences among users may all be relevant. Because of this heterogeneity, not all users will have the same position in relation to a specific technology. For some users, the room for maneuvering will be great; for others, it will be very slight. Feminist sociologists thus emphasize the diversity of users and encourage scholars to pay attention to differences in power relations among the actors involved in the development of technology.

To capture the diversity of users and the power relations between users and other actors in technological development, feminist sociologists have differentiated "end users," "lay end users," and "implicated actors."

End users are "those individuals and groups who are affected downstream by products of technological innovation" (Casper and Clarke 1998). The term "lay end users" was introduced to highlight some end users' relative exclusion from expert discourse (Saetnan et al. 2000: 16). Implicated actors are "those silent or not present but affected by the action" (Clarke 1998: 267). And there are two categories of implicated actors: "those not physically present but who are discursively constructed and targeted by others" and "those who are physically present but who are generally silenced/ignored/made invisible by those in power" (Clarke, forthcoming). All three terms reflect the long-standing feminist concern with the potential problematic consequences of technologies for women and include an explicit political agenda: the aim of feminist studies is to increase women's autonomy and their influence on technological development. A detailed understanding of how women as "end users" or "implicated actors" matter in technological development may provide information that will be useful in the empowerment of women or of spokespersons for them, such as social movements and consumer groups.

The concept of the implicated actor also reflects a critical departure from actor-network approaches in technology studies. Feminists have criticized the sociology of technology, particularly actor-network theory, for the almost exclusive attention it gives to experts and producers and for the preference it gives to design and innovation in understanding socio-technical change. This "executive approach" pays less attention to non-standard positions, including women's voices (Star 1991; Clarke and Montini 1993: 45; Clarke 1998: 267). Moreover, the "executive approach" implicitly assumes a specific type of power relations between users and designers in which designers are represented as powerful and users as disempowered relative to the experts. Feminist sociologists suggest that the distribution of power among the multiple actors involved in socio-technical networks should be approached as an empirical question (Lie and Sørensen 1996: 4, 5; Clarke 1998: 267; Oudshoorn et al., forthcoming). Thus, the notion of the implicated actor was introduced to avoid silencing invisible actors and actants and to include power relations explicitly in the analysis of user-expert relations.

Another important word in the feminist vocabulary is "cyborg." Donna Haraway was the first to use this word to describe how by the late twentieth century humans had become so thoroughly and radically merged and fused with technologies that the boundaries between the human and the technological are no longer impermeable. The cyborg implies a very specific configuration of user-technology relations in which the user emerges as a hybrid of machine and organisms in fiction and as lived experience. Most important, Haraway introduced the cyborg figure as a politicized entity. Cyborg analyses aim to go beyond the deconstruction of technological discourses. On page 149 of her "Cyborg Manifesto" (1985), Haraway invites us to "question that which is taken as 'natural' and 'normal' in hierarchic social relations." Haraway writes of cyborgs not to celebrate the fusion of humans and technology, but to subvert and displace meanings in order to create alternative views, languages, and practices of technosciences and hybrid subjects. In the 1990s, the concept of the cyborg resulted in an extensive body of literature that described the constitution and transformation of physical bodies and identities through technological practices.

■ Semiotic Approaches to Users: Configuration and Script

An important new approach to user-technology relations was introduced by STS scholars who extended semiotics, the study of how meanings are built, from signs to things. The concept of "configuring the user" is central to this approach. Exploring the metaphor of machine as text, Steve Woolgar (1991: 60) introduced the notion of the user as reader to emphasize the interpretive flexibility of technological objects and the processes that delimit this flexibility. Although the interpretive flexibility of technologies and questions concerning the closure or stabilization of technology had already been addressed in the SCOT approach, Woolgar focused on the design processes that delimit the flexibility of machines rather than on the negotiations between relevant social groups. He suggested that how users "read" machines is constrained because the design and the production of machines entails a process of configuring the user. For Woolgar, "configuring" is the process of "defining the identity of putative users, and setting constraints upon their likely future actions" (ibid.: 59). He describes the testing of a new range of microcomputers as "a struggle to configure (that is to define, enable, and constrain) the user," a struggle that results in "a machine that encourages only specific forms of access and use" (ibid.: 69, 89). In this approach, the testing phase of a technology is portrayed as an important location in which to study the co-construction of technologies and users. In contrast to the approaches discussed thus far, this semiotic approach draws attention to users as represented by designers rather than to users as individuals or groups involved or implicated in technological innovation.

In recent debates, the notion of the configuration of users by designers has been extended to better capture the complexities of designer-user relations. Several authors criticized Woolgar for describing-configuration as a one-way process in which the power to shape technological development is attributed only to experts in design organizations. For example, Mackay et al. (2000: 752) suggested that "designers configure users, but designers in turn, are configured by both users and their own organizations," and that this is increasingly the case in situations where designer-user relations are formalized by contractual arrangements (ibid.: 744). The capacity of designers to configure users can be further constrained by powerful groups within organizations that direct design projects. In large organizations, designers usually have to follow specific organizational methods or procedures that constrain design practices (ibid.: 741, 742, 744; Oudshoorn et al. 2003). In many companies in the information and communication technologies sector, for example, designers are allowed to test prototypes of new products only among people who work in the organization. In this highly competitive sector, companies are reluctant to test new products among wider groups of users for fear that other firms will become aware of their plans at an early phase of product development (European Commission 1998: 22; Oudshoorn et al. 2003).

Another criticism and extension of the configuration approach was introduced by scholars who questioned who was doing the configuration work. In Woolgar's studies, configuration work was restricted to the activities of actors within the company who produced the computers. Several authors broadened this rather narrow view of configuration to include other actors and to draw attention to the configuration work carried out by journalists, public-sector agencies, policy makers, and social movements acting as spokespersons for users (van Kammen 2000a; van Kammen, this volume; Epstein, this volume; Parthasarathy, this volume; Oudshoorn 1999; Rommes 2002). Other scholars attempted to broaden the scope of the analysis by including the agency of users. Whereas Woolgar explored the metaphor of machine and text to highlight "encoding," thus focusing attention on the work performed by the producers of texts and machines, a more symmetrical use of the metaphor requires that we also focus on the processes of "decoding," the work done by readers and users to interpret texts and machines (Mackay et al. 2000: 739, 750, 752). A similar criticism of the asymmetry of Woolgar's work was voiced by scholars who had adopted domestication approaches to technology.

A second central notion in the semiotic approaches to user-technology relations is the concept of script. Madeleine Akrich and Bruno Latour, in theorizing relationships between users and technology, use this term to describe the obduracy of objects. The concept of script tries to capture how technological objects enable or constrain human relations as well as relationships between people and things. Comparing technologies to film, Akrich (1992: 208) suggested that "like a film script, technical objects define a framework of action together with the actors and the space in which they are supposed to act." To explain how scripts of technological objects emerge, she drew attention to the design of technologies. Akrich suggested that in the design phase technologists anticipate the interests, skills, motives, and behavior of future users. Subsequently, these representations of users become materialized into the design of the new product. As a result, technologies contain a script (or scenario): they attribute and delegate specific competencies, actions, and responsibilities to users and technological artifacts. Technological objects may thus create new "geographies of responsibilities" or transform or reinforce existing ones (ibid.: 207, 208). Rooted in actor-network theory, Akrich and Latour's work challenges social constructivist approaches in which only people are given the status of actors. The script approach aims to describe how technical objects "participate in building heterogeneous networks that bring together actants of all types and sizes, whether humans or nonhumans" (ibid.: 206).

In the 1990s, feminist scholars extended the script approach to include the gender aspects of technological innovation. Adopting the view that technological innovation requires a renegotiation of gender relations and an articulation and performance of gender identities, Dutch and Norwegian feminists introduced the concept of genderscript to capture all the work involved in the inscription and de-inscription of representations of masculinities and femininities in technological artifacts (Berg and Lie 1993; Hubak 1996; van Oost 1995; van Oost, this volume; Oudshoorn 1996; Oudshoorn et al. 2003; Oudshoorn et al., forthcoming; Rommes et al. 1999; Spilkner and Sørensen 2000). This scholarship emphasizes the importance of studying the inscription of gender into artifacts to improve our understanding of how technologies invite or inhibit specific performances of gender identities and relations. Technologies are represented as objects of identity projects—objects that may stabilize or de-stabilize hegemonic representations of gender (Oudshoorn, this volume; Saetnan et al. 2000). Equally important, the genderscript approach drastically redefines the exclusion of specific groups of people from technological domains and activities. Whereas policy makers and researchers have defined the problem largely in terms of deficiencies of users, genderscript studies draw attention to the design of technologies (Oudshoorn 1996; Oudshoorn et al., forthcoming; Rommes et al. 1999; Rommes 2002). These studies make visible how specific practices of configuring the user may lead to the exclusion of specific users.

At first glance, the script approach seems to be very similar to Woolgar's approach of configuring the user, since both approaches are concerned with understanding how designers inscribe their views of users and use in technological objects. A closer look, however, reveals important differences. Although both approaches deal with technological objects and designers, the script approach makes users more visible as active participants in technological development. Akrich in particular is very much aware that a focus on how technological objects constrain the ways in which people relate to things and to one another can be easily misunderstood as a technological determinist view that represents designers as active and users as passive. To avoid this misreading, she emphasizes the reciprocal relationship between objects and subjects and explicitly addresses the question of the agency of users (Akrich 1992: 207). Although technological objects can define the relationships between human and nonhuman actors, Akrich suggests that "this geography is open to question and may be resisted" (ibid.). To avoid technological determinism, Akrich urges us to analyze the negotiations between designers and users and concludes that "we cannot be satisfied methodologically with the designer's or user's point of view alone. Instead we have to go back and forth continually between the designer and the user, between the designer's projected users and the real users, between the world inscribed in the object and the world described by its displacement" (ibid.: 209).

To further capture the active role of users in shaping their relationships to technical objects, Akrich and Latour have introduced the concepts of subscription, de-inscription, and antiprogram. "Antiprogram" refers to the users' program of action that is in conflict with the designers' program (or vice versa). "Subscription" or "de-inscription" is used to describe the reactions of human (and nonhuman) actors to "what is prescribed and proscribed to them" and refers respectively to the extent to which they underwrite or reject and renegotiate the prescriptions (Akrich and Latour 1992: 261). In contrast to Woolgar's work on configuring the user, script analyses thus conceptualize both designers and users as active agents in the development of technology. However, compared to domestication theory, the script approach gives more weight to the world of designers and technological objects. The world of users, particularly the cultural and social processes that facilitate or constrain the emergence of users' antiprograms, remains largely unexplored by actor-network approaches. More recently, this imbalance has been repaired to an extent by the work of scholars who have extended actor-network theory to include the study of "subject networks." These studies aim to understand the "attachment" between people and things, particularly but not exclusively between disabled people and assistive technologies, and to explore how technologies work to articulate subjectivities (Callon and Rabeharisoa 1999; Moser 2000; Moser and Law 1998, 2001). This scholarship conceptualizes subjects in the same way as actor-network theorists previously approached objects. Subject positions such as disability and ability are constituted as effects of actor networks and hybrid collectives.

◼ Cultural and Media Studies: Consumption and Domestication

In contrast to the approaches to user-technology relations we have discussed thus far, scholars in the fields of cultural and media studies acknowledged the importance of studying users from the very beginning. Whereas historians and sociologists of technology have chosen technology as their major topic of analysis, those who do cultural and media studies have focused primarily on users and consumers. Their central thesis is that technologies must be culturally appropriated to become functional. This scholarship draws inspiration from Bourdieu's (1984) suggestion that consumption has become more important in the political economy of late modernity. Consequently, human relations and identities are increasingly defined in relation to consumption rather than production. In his study of differences in consumption patterns among social classes, Bourdieu defined consumption as a cultural and material activity and argued that the cultural appropriation of consumer goods depends on the "cultural capital" of people (ibid.). This view can be traced back to the tradition of the anthropological study of material culture, most notably the work of Mary Douglas and

Baron Isherwood (1979). Among the first to criticize the view (then dominant among consumption theorists) that consumption is merely an economic activity, they suggested that consumption is always a cultural as well as an economic phenomenon (Lury 1996: 10). Describing the use of consumer goods in ritual processes, they defined consumer culture as a specific form of material culture, and they conceptualized the circulation of material things as a system of symbolic exchange. This scholarship articulates the importance of the sign value rather than the utility value of things. From this perspective, material things can act as sources and markers of social relations and can shape and create social identities (Lury 1996: 10, 12, 14; Douglas and Isherwood 1979; McKracken 1988; Appadurai 1986).

Feminist historians have also been important actors in signaling the relevance of studying consumption rather than production (McGaw 1982). Feminists have long been aware of the conventional association and structural relations of women with consumption as a consequence of their role in the household and as objects in the commodity-exchange system (de Grazia 1996: 7). Whereas early feminist studies focused on the (negative) consequences of mass consumption for women, more recent studies address the question of whether women have been empowered by access to consumer goods. They conceptualize consumption as a site for the performance of gender and other identities. The notion of consumption as a status and identity project was elaborated further by Jean Baudrillard (1988), who criticizes the view that the needs of consumers are dictated, manipulated, and fully controlled by the modern capitalist marketplace and by producers. Theodor Adorno, Herbert Marcuse, and Max Horkheimer of the Frankfurt School had argued that the expansion of the production of consumer goods throughout the twentieth century had resulted in an increase in ideological control and manipulation by the "culture industries" (Adorno 1991; Horkheimer and Adorno 1979; Marcuse 1964). Since the 1970s, this view of consumption as manipulation had resulted in a literature dominated by studies oriented toward production and marketing—studies that highlighted big companies and advertising agencies as the forces driving consumption. In these studies, consumption was characterized as a passive and adaptive process and consumers are represented as the anonymous buyers and victims of mass production. In contrast, Baudrillard emphasized the mutual dependencies between production and consumption and suggested that consumers are not passive victims but active agents in shaping consumption, social relations, and identities.

Cultural and media studies also emphasize the creative freedom of users to "make culture" in the practice of consumption as well as their dependence on the cultural industries, not because they control consumers but because they provide the means and the conditions of cultural creativity (Storey 1999: xi). This scholarship portrays consumers as "cultural experts" who appropriate consumer goods to perform identities, which may transgress established social divisions (du Gay et al. 1997: 104; Chambers 1985).

Semiotic approaches to analyzing user-technology relations also came to the fore in cultural and media studies. Stuart Hall, one of the leading scholars in this field, introduced the "encoding/decoding" model of media consumption (Hall 1973), which aims to capture both the structuring role of the media in "setting agendas and providing cultural categories and frameworks" and the notion of the "active viewer, who makes meaning from signs and symbols that the media provide" (Morley 1995: 300). Since the 1980s, the symbolic and communicative character of consumption has been studied extensively by scholars in the fields of cultural and media studies. Consumption fulfills a wide range of social and personal aims and serves to articulate who we are or who we would like to be; it may provide symbolic means of creating and establishing friendship and celebrating success; it may serve to produce certain lifestyles; it may provide the material for daydreams; it may be used to articulate social difference and social distinctions (Bocock 1993; du Gay et al. 1997; Lie and Sørensen 1996; Mackay 1997; Miller 1995; Storey 1999). Cultural and media studies thus articulate a perspective on user-technology relations that emphasizes the role of technological objects in creating and shaping-social identities, social life, and culture at large.

Roger Silverstone coined the term "domestication" to describe how the integration of technological objects into daily life involves "a taming of the wild and a cultivation of the tame." New technologies have to be transformed from "unfamiliar, exciting, and possible threatening things" into familiar objects embedded in the culture of society and the practices and routines of everyday life (Silverstone and Hirsch 1992; Lie and Sørensen 1996). Domestication processes include symbolic work, in which people create symbolic meanings of artifacts and adopt or transform the meanings inscribed in the technology; practical work, in which users develop a pattern of use to integrate artifacts into their daily routines; and cognitive work, which includes learning about artifacts (Lie and Sørensen 1996: 10; Sørensen et al. 2000). In this approach, domestication is defined as a dual process in which both technical objects and people may change. The use of technological objects may change the form and the practical and symbolic functions of artifacts, and it may enable or constrain performances of identities and negotiations of status and social positions (Silverstone et al. 1989; Lie and Sørensen 1996). The notion of domestication also reflects a preference for studying the use of technology in a specific location: the home. British scholars in this tradition have largely restricted their analyses to the household and the politics of family life (Silverstone 1989, 1992). In their work, processes of domestication are understood in terms of the "dynamics of the household's moral economy" (Silverstone, Hirsch, and Morley 1992). More recently, Norwegian scholars have extended the scope of research to other domains. Merete Lie and Knut Sørensen (1996: 13, 17) argue that the domestication of technical objects has been too easily associated with the "private sector" (meaning the home). Various chapters in the volume edited by Lie and Sørensen show how similar processes are taking-place in work, in leisure, and within subcultures.

Domestication approaches have enriched our understanding of user-technology relations by elaborating the processes involved in consumption. In *Consuming Technologies,* Roger Silverstone and his colleagues specify four phases of domestication: appropriation, objectification, incorporation, and conversion. Appropriation occurs when a technical product or service is sold and individuals or households become its owners (Silverstone et al. 1992: 21). In objectification, processes of display reveal the norms and principles of the "household's sense of itself and its place in the world" (ibid.: 22). Incorporation occurs when technological objects are used in and incorporated into the routines of daily life. "Conversion" is used to describe the processes in which the use of technological objects shape relationships between users and people outside the household (ibid.: 25). In this process, artifacts become tools for making status claims and for expressing a specific lifestyle to neighbors, colleagues, family, and friends (Silverstone and Haddon 1996: 46).

Although at first sight "domestication" and "decoding" or "de-inscription" may seem synonymous, there is an important difference. By specifying the processes involved in the diffusion and the use of technology, domestication approaches take the dynamics of the world of users as their point of departure. The concepts of decoding and de-inscription, on the other hand, give priority to the design context in order to understand the emergence of user-technology relations. Domestication approaches thus emphasize the complex cultural dynamics in which users appropriate technologies (ibid.: 52). This contrasts with semiotic approaches that tend to define the user as an isolated individual whose relationship to technology is restricted to technical interactions with artifacts. As Silverstone and Haddon suggest, a focus on how designers configure the user runs the risk of reifying the innovator's conceptions of users. In contrast, domestication approaches conceptualize the user as a part of a much broader set of relations than user-machine interactions, including social, cultural, and economic aspects. By employing cultural approaches to understand user-technology relations, this scholarship aims to go beyond a rhetoric of designers' being in control. Semiotic approaches tend to reinforce the view that technological innovation and diffusion are successful only if designers are able to control the future actions of users. Although semiotic approaches have introduced notions that are useful in understanding the worlds of designers and users, "script" and "configuring the user" conceptualize the successes and failures of technologies mainly in terms of the extent to which designers adequately anticipate users' skills and behavior. In this view, users tend to be degraded to objects of innovators' strategies. The semiotic approaches have therefore been criticized for staying too close to the old linear model of technological innovation and diffusion, which prioritizes the agency of designers and producers over the agency of users and other actors involved in technological innovation (Oudshoorn 1999). Even the concept of antiprogram, introduced by Akrich and Latour to describe how users may try to counter the original intentions of the design of the artifact, remains within the rhetoric of designer's control (Sørensen 1994: 5). The only option available to the user seems to be to adopt or to reject the designers' intended use and meaning of technological objects. These approaches are inadequate to understand the full dynamics of technological innovation where users invent completely new uses and meanings of technologies or where users are actively involved in the design of technologies.

Most important, cultural and media studies inspire us to transcend the artificial divide between design and use. This scholarship has drastically reconceptualized the traditional distinction between production and consumption by re-introducing Karl Marx's claim that the process of production is not complete until users have defined the uses, meanings, and significance of the technology: "Consumption is production." They describe design and domestication as "the two sides of the innovation coin" (Lie and Sørensen 1996: 10).

"Touch Someone": The Telephone Industry Discovers Sociability

Reading 10

Claude S. Fischer

The familiar refrain, "Reach out, reach out and touch someone," has been part of American Telephone and Telegraph's (AT&T's) campaign urging use of the telephone for personal conversations. Yet, the telephone industry did not always promote such sociability; for decades it was more likely to discourage it. The industry's "discovery" of sociability illustrates how structural and cultural constraints interact with public demand to shape the diffusion of a technology. While historians have corrected simplistic notions of "autonomous technology" in showing how technologies are produced, we know much less about how consumers use technologies. We too often take those uses (especially of consumer products) for granted, as if they were straightforwardly derived from the nature of the technology or dictated by its creators.[1]

In the case of the telephone, the initial uses suggested by its promoters were determined by—in addition to technical and economic considerations—its cultural heritage: specifically, practical uses in common with the telegraph. Subscribers nevertheless persisted in using the telephone for "trivial gossip." In the 1920s, the telephone industry shifted from resisting to endorsing such sociability, responding, at least partly, to consumers' insistent and innovative uses of the technology for personal conversation. After summarizing telephone history to 1940, this article will describe the changes in the uses that telephone promoters advertised and the changes in their attitudes toward sociability; it will then explore explanations for these changes.[2]

---

[1] See C. S. Fischer, "Studying Technology and Social Life," pp. 284–301 in *High Technology, Space, and Society: Emerging Trends,* ed. M. Castells (Beverly Hills, Calif., 1985). For a recent example of a study looking at consumers and sales, see M. Rose, "Urban Environments and Technological Innovation: Energy Choices in Denver and Kansas City, 1900–1940," *Technology and Culture* 25 (July 1984): 503–39.

[2] The primary sources used here include telephone and advertising industry journals; internal telephone company reports, correspondence, collections of advertisements, and other documents, primarily from AT&T and Pacific Telephone (PT&T); privately published memoirs and corporate histories; government censuses, investigations, and research studies; and several interviews, conducted by John Chan, with retired telephone company employees who had worked in marketing. The archives used most are the AT&T Historical Archives, New York (abbreviated hereafter as AT&T ARCH), and the Pioneer Telephone Museum, San Francisco (SF PION MU), with some material from the Museum of Independent Telephony, Abilene (MU IND TEL); Bell Canada Historical, Montreal (BELL CAN HIST); Illinois Bell Information Center, Chicago (ILL BELL INFO); and the N. W. Ayer Collection of Advertisements and the Warshaw Collection of Business Americana, National Museum of American History, Smithsonian Institution, Washington, D.C. A bibliography on the social history of the telephone is unusually short, especially in comparison with those on later technologies such as the automobile and television. There are industrial and corporate histories, but the consumer side is largely untouched. For some basic sources, see J. W. Stehman, *The Financial History of the American Telephone and Telegraph Company* (Boston, 1925); A. N. Holcombe, *Public Ownership of Telephones on the Continent of Europe* (Cambridge, Mass., 1911); H. B. MacMeal, *The Story of Independent Telephony* (Chicago: Independent Pioneer Telephone Association, 1934); J. L. Walsh, *Connecticut Pioneers in Telephony* (New Haven, Conn.: Morris F. Tyler Chapter of the Telephone Pioneers of America, 1950); J. Brooks, *Telephone: The First Hundred Years* (New York, 1976); A. Hibbard, *Hello-Goodbye: My Story of Telephone Pioneering* (Chicago, 1941); Robert Collins, *A Voice from Afar: The History of Telecommunications in Canada* (Toronto, 1977); R. L. Mahon, "The Telephone in Chicago," ILL BELL INFO, MS, ca. 1955; J. C. Rippey, *Goodbye, Central; Hello, World: A Centennial History of Northwestern Bell* (Omaha, Nebr.: Northwestern Bell, 1975); G. W. Brock, *The Telecommunications Industry: The Dynamics of Market Structure* (Cambridge, Mass., 1981); I. de S. Pool, *Forecasting the Telephone* (Norwood, N.J., 1983); R. W. Garnet, *The Telephone Enterprise: The Evolution of the Bell System's*

■ A Brief History of the Telephone

Within about two years of A. G. Bell's patent award in 1876, there were roughly 10,000 Bell telephones in the United States and fierce patent disputes over them, battles from which the Bell Company (later to be AT&T) emerged a victorious monopoly. Its local franchisees' subscriber lists grew rapidly and the number of telephones tripled between 1880 and 1884. Growth slowed during the next several years, but the number of instruments totaled 266,000 by 1893.[3] (See table 1.)

As long-distance communication, telephony quickly threatened telegraphy. Indeed, in settling its early patent battle with Western Union, Bell gave financial concessions to Western Union as compensation for loss of business. As local communication, telephony quickly overwhelmed nascent efforts to establish signaling exchange systems (except for stock tickers).

During Bell's monopoly, before 1894, telephone service consisted basically of an individual line for which a customer paid an annual flat fee allowing unlimited calls within the exchange area. Fees varied widely, particularly by size of exchange. Bell rates dropped in the mid-1890s, perhaps in anticipation of forthcoming competition. In 1895, Bell's average residential rate was $4.66 a month (13 percent of an average worker's

TABLE 1
TELEPHONE DEVELOPMENT, 1880–1940

	Number of Telephones	Telephones per 1,000 People	Percentage in Bell System	Percentage Independent Connected to Bell	Percentage Residential Connected to Bell
1880	54,000	1	100	0	. . .
1885	156,000	3	100	0	. . .
1890	228,000	4	100	0	. . .
1895	340,000	5	91	0	. . .
1900	1,356,000	18	62	1	. . .
1905	4,127,000	49	55	6	. . .
1910	7,635,000	82	52	26	. . .
1915	10,524,000	104	57	30	. . .
1920	13,273,000	123	66	29	68
1925	16,875,000	145	75	24	67
1930	20,103,000	163	80	20	65
1935	17,424,000	136	82	18	63
1940	21,928,000	165	84	16	65
1980	180,000,000	790	81	19	74

SOURCES—U.S. Bureau of the Census, *Historical Statistics of the United States,* Bicentennial Ed., pt. 2 (Washington, D.C., 1975), pp. 783–84; and U.S. Bureau of the Census, *Statistical Abstract of the United States 1982–83* (Washington, D.C., 1984), p. 557.

Horizontal Structure, 1876–1909 (Baltimore, 1985); R. A. Atwood, "Telephony and Its Cultural Meanings in Southeastern Iowa, 1900–1917" (Ph.D. diss., University of Iowa, 1984); Lana Fay Rakow, "Gender, Communication, and the Technology: A Case Study of Women and the Telephone" (Ph.D. diss., University of Illinois at Urbana-Champaign, 1987); and I. de S. Pool, ed., *The Social Impact of the Telephone* (Cambridge, Mass., 1977). (Note that AT&T, Bell, and similar corporate names refer, of course, to these companies—or their direct ancestors—up to the U.S. industry reorganization of January 1, 1984.)

[3]Statistics from AT&T, *Events in Telecommunications History* (New York: AT&T, 1979), p. 6; U.S. Bureau of the Census (BOC), *Historical Statistics of the United States*, Bicentennial Ed., pt. 2 (Washington, D.C., 1975), pp. 783–84.

monthly wages). Rates remained high, especially in the larger cities (the 1894 Manhattan rate for a two-party line was $10.41 a month).[4]

On expiration of the original patents in 1893–94, thousands of new telephone vendors, ranging from commercial operations to small cooperative systems, sprang up. Although they typically served areas that Bell had ignored, occasional head-to-head competition drove costs down and spurred rapid diffusion: almost a nine-fold increase in telephones per capita between 1893 and 1902, as compared to less than a twofold increase in the prior nine years.[5]

Bell responded fiercely to the competition, engaging in price wars, political confrontations, and other aggressive tactics. It also tried to reach less affluent customers with cheaper party lines, coin-box telephones, and "measured service" (charging by the call). Still, Bell lost at least half the market by 1907. Then, a new management under Theodore N. Vail, the most influential figure in telephone history, changed strategies. Instead of reckless, preemptive expansion and price competition, AT&T bought out competitors where it could and ceded territories where it was losing. With tighter fiscal control, and facing capital uncertainties as well, AT&T's rate of expansion declined.[6] Meanwhile, the "independents" could not expand much beyond their small-town bases, partly because they were unable to build their own long-distance lines and were cut off from Bell-controlled New York City. Many were not competitive because they were poorly financed and provided poor service. Others accepted or even solicited buyouts from AT&T or its allies. By 1912, the Bell System had regained an additional 6 percent of the market.

During this competitive era, the industry offered residential customers a variety of economical party-line plans. Bell's average residential rate in 1909 was just under two dollars a month (about 4 percent of average wages).[7] How much territory the local exchange covered and what services were provided—for example, nighttime operators—varied greatly, but costs dropped and subscriber lists grew considerably. These basic rates changed little until World War II (although long-distance charges dropped).

In the face of impending federal antitrust moves, AT&T agreed in late 1913 to formalize its budding accommodation with the independents. Over several years, local telephone service was divided into regulated geographic monopolies. The modern U.S. telephone system—predominantly Bell local service and exclusively Bell long-distance service—was essentially fixed from the early 1920s to 1984.

The astronomical growth in the number of telephones during the pre-Vail era (a compound annual rate of 23 percent per capita from 1893 to 1907) became simply healthy growth (4 percent between 1907 and 1929). The system was consolidated and technically improved, and, by 1929, 42 percent of all households had telephones. That figure shrank during the Depression to 31 percent in 1933 but rebounded to 37 percent of all households in 1940.

■ Sales Strategies

The telephone industry believed, as President Vail testified in 1909, that the "public had to be educated . . . to the necessity and advantage of the telephone."[8] And Bell saluted itself on its success in an advertisement entitled "Blazing the Way": Bell "had to invent the business uses of the telephone and convince people that

---

[4]Rates are reported in scattered places. For these figures, see BOC, *Telephones and Telegraphs 1902*, Special Reports, Department of Commerce and Labor (Washington, D.C., 1906), p. 53; and *1909 Annual Report of AT&T* (New York, 1910), p. 28. Wage data are from *Historical Statistics* (n. 3 above), tables D735–38.

[5]BOC, *Telephones, 1902* (n. 4 above); Federal Communications Commission (FCC), *Proposed Report: Telephone Investigation* (Washington, D.C., 1938), p. 147. AT&T has always officially challenged this interpretation; see, e.g., *1909 Annual Report of AT&T*, pp. 26–28.

[6]See, e.g., *Annual Report of AT&T,* 1907–10; and FCC, *Proposed Report* (n. 5 above), pp. 153–154. On making deals with competitors, see, e.g., Rippey (n. 2 above), pp. 143ff.

[7]*1909 Annual Report of AT&T*, p. 28. Charges for minimal, urban, four-party lines ranged from $3.00 a month in New York (about 6 percent of the average manufacturing employee's monthly wages) to $1.50 in Los Angeles (about 3 percent of wages) and much less in small places with mutual systems; see BOC, *Telephones and Telegraphs and Municipal Electric Fire-Alarm and Police-Patrol Signaling Systems, 1912* (Washington, D.C., 1915); and *Historical Statistics* (n. 3 above), table D740.

[8]Testimony on December 9, 1909, in State of New York, *Report of the Committee of the Senate and Assembly Appointed to Investigate Telephone and Telegraph Companies* (Albany, 1910), p. 398.

they were uses. . . . [Bell] built up the telephone habit in cities like New York and Chicago. . . . It has from the start created the need of the telephone and then supplied it."[9]

"Educating the public" typically meant advertising, face-to-face solicitations, and public relations. In the early years, these efforts included informational campaigns, such as publicizing the existence of the telephone, showing people how to use it, and encouraging courteous conversation on the line.[10] Once the threat of nationalization became serious, "institutional" advertising and publicity encouraged voters to feel warmly toward the industry.[11]

As to getting paying customers, the first question vendors had to ask was, Of what use is this machine? The answer was not self-evident.

For roughly the first twenty-five years, sales campaigns largely employed flyers, simple informational notices in newspapers, "news" stories supplied to friendly editors (many of whom received free service or were partners in telephony), public demonstrations, and personal solicitations of businessmen. As to uses, salesmen typically stressed those that extended applications of telegraph signaling. For example, an 1878 circular in New Haven—where the first exchange was set up—stated that "your wife may order your dinner, a hack, your family physician, etc., all by Telephone without leaving the house or trusting servants or messengers to do it." (It got almost no response.)[12] In these uses, the telephone directly competed with—and decisively defeated—attempts to create telegraph exchanges that enabled subscribers to signal for services and also efforts to employ printing telegraphs as a sort of "electronic mail" system.[13]

In this era and for some years later, the telephone marketers sought new uses to add to these telegraphic applications. They offered special services over the telephone, such as weather reports, concerts, sports results, and train arrivals. For decades, vendors cast about for novel applications: broadcasting news, sports, and music, night watchman call-in services, and the like. Industry magazines eagerly printed stories about the telephone being used to sell products, alert firefighters about forest blazes, lullaby a baby to sleep, and get out voters on election day. And yet, industry men often attributed weak demand to not having taught the customer "what to do with his telephone."[14]

---

[9]Ayer Collection of AT&T Advertisements, Collection of Business Americana, National Museum of American History, Smithsonian Institution.

[10]See, e.g., *Pacific Telephone Magazine* (PT&T employee magazine, hereafter PAC TEL MAG), 1907–40, passim; 1914 advertisements in SF PION MU folder labeled "Advertising"; MU IND TEL "Scrapbook" of Southern Indiana Telephone Company clippings; advertisements in directories of the day; "Educating the Public to the Proper Use of the Telephone," *Telephony* 64 (June 21, 1913): 32–33; "Swearing over the Telephone," *Telephony* 9 (1905): 418; and "Advertising and Publicity—1906–1910," box 1317, AT&T ARCH.

[11]On AT&T's institutional advertising, see R. Marchand, "Creating the Corporate Soul: The Origins of Corporate Image Advertising in America" (paper presented to the Organization of American Historians, 1980), and N. L. Griese, "AT&T: 1908 Origins of the Nation's Oldest Continuous Institutional Advertising Campaign," *Journal of Advertising* 6 (Summer 1977): 18–24. FCC, *Proposed Report* (n. 5 above), has a chapter on "Public Relations"; see also N. R. Danielian, *AT&T: The Story of Industrial Conquest* (New York, 1939), chap. 13. For a defense of AT&T public relations, see A. W. Page, *The Bell Telephone System* (New York, 1941). Among the publicity efforts along these lines were "free" stories, subsidies of the press, and courting of reporters and politicians (documented in AT&T ARCH). In one comical case, AT&T frantically and apparently unsuccessfully tried in 1920 to pressure Hal Roach to cut out from a Harold Lloyd film he was producing a burlesque scene of central exchange hysteria (see folder "Correspondence—E. S. Wilson, V.P., AT&T," SF PION MU).

[12]Walsh (n. 2 above), p. 47.

[13]S. Schmidt, "The Telephone Comes to Pittsburgh" (master's thesis, University of Pittsburgh, 1948); Pool, *Forecasting* (n. 2 above), p. 30; D. Goodman, "Early Electrical Communications and the City: Applications of the Telegraph in Nineteenth-Century Urban America" (unpub. paper, Department of Social Sciences, Carnegie-Mellon University, n.d., courtesy of Joel Tarr); and "Telephone History of Dundee, Ontario," City File, BELL CAN HIST.

[14]On special services and broadcasting, see Walsh (n. 2 above), p. 206; S. H. Aronson, "Bell's Electrical Toy: What's the Use? The Sociology of Early Telephone Usage," pp. 15–39, and I. de S. Pool et al., "Foresight and Hindsight: The Case of the Telephone," pp. 127–58, both in Pool, ed., *Social Impact* (n. 2 above); "Broadening the Possible Market," *Printers' Ink* 74 (March 9, 1911): 20; G. O. Steel, "Advertising the Telephone," *Printers' Ink* 51 (April 12, 1905): 14–17; and F. P. Valentine, "Some Phases of the Commercial Job," *Bell Telephone Quarterly* 5 (January 1926): 34–43. For illustrations of uses, see, e.g., PAC TEL MAG (October 1907), p. 6, (January 1910), p. 9, (December 1912), p. 23, and (October 1920), p. 44; and the independent magazine, *Telephony*. E.g., the index to vol. 71 (1916) of *Telephony* lists the following under "Telephone, novel uses of": "degree conferred by telephone, dispatching tugs in harbor service, gauging water by telephone, telephoning in an aeroplane." On complaints about not having taught the public, see the quotation from H. B. Young, ca. 1929, pp. 91, 100 in "Publicity Conferences—Bell System—1921–34," box 1310, AT&T ARCH, but similar comments appear in earlier years, as well as positive claims, such as Vail's in 1909.

In the first two decades of the 20th century, telephone advertising became more professionally "modern."[15] AT&T employed a Boston agency to dispense "free publicity" and later brought its chief, J. D. Ellsworth, into the company. It began national advertising campaigns and supplied local Bell companies with copy for their regional presses. Some of the advertising was implicitly competitive (e.g., stressing that Bell had long-distance service), and much of it was institutional, directed toward shaping a favorable public opinion about the Bell System. Advertisements for selling service employed drawings, slogans, and texts designed to make the uses of the telephone—not just the technology—attractive. (The amount and kind of advertising fluctuated, especially in the Bell System, in response to competition, available supplies, and political concerns.)[16]

From roughly 1900 to World War I, Bell's publicity agency advertised uses of the telephone by planting newspaper "stories" on telephones in farm life, in the church, in hotels, and the like.[17] The national advertisements, beginning around 1910, addressed mostly businessmen. They stressed that the telephone was impressive to customers and saved time, both at work and at home, and often noted the telephone's convenience for planning and for keeping in touch with the office during vacations.

A second major theme was household management. A 1910 series, for example, presented detailed suggestions: Subscribers could telephone dressmakers, florists, theaters, inns, rental agents, coal dealers, schools, and the like. Other uses were suggested, too, such as conveying messages of moderate urgency (a businessman calling home to say that he will be late, calling a plumber), and conveying invitations (to an impromptu party, for a fourth at bridge).

Sociability themes ("visiting" kin by telephone, calling home from a business trip, and keeping "In Touch with Friends and Relatives") appeared, but they were relatively rare and almost always suggested sending a message such as an invitation or news of safe arrival rather than having a conversation. A few advertisements also pointed out the modernity of the telephone ("It's up to the times!"). But the major uses suggested in early telephone advertising were for business and household management; sociability was rarely advised.[18]

With the decline of competition and the increase in regulation during the 1910s, Bell stressed public relations even more and pressed local companies to follow suit. AT&T increasingly left advertising basic services and uses to its subsidiaries, although much of the copy still originated in New York, and the volume of such advertising declined. Material from Pacific Telephone and Telegraph (PT&T), apparently a major advertiser among the Bell companies, indicates the substance of "use" advertising during that era.[19]

PT&T advertisements for 1914 and 1915 include, aside from informational notices and general paeans to the telephone, a few suggestions for businessmen (e.g., "You fishermen who feel these warm days of Spring luring you to your favorite stream. . . . You can adjust affairs before leaving, ascertain the condition of streams, secure accommodations, and always be in touch with business and home"). Several advertisements mention the home or women, such as those suggesting that extension telephones add to safety and those encouraging

---

[15]The following discussion draws largely from examination of advertisement collections at the archives listed in n. 2. Space does not permit more than a few examples of hundreds of advertisements in the sources. See esp. at AT&T ARCH, files labeled "Advertising and Publicity"; at SF PION MU, folders labeled "Advertising" and "Publicity Bureau"; at BELL CAN HIST, "Scrapbooks"; at ILL BELL INFO, "AT&T Advertising" and microfilm 384B, "Adver."; and at the Ayer Collection (n. 9 above), the AT&T series.

[16]For explicit discussions, see Mahon (n. 2 above), e.g., pp. 79, 89; Publicity Vice-President A. W. Page's comments in "Bell System General Commercial Conference, 1930," microfilm 368B, ILL BELL INFO; and comments by Commercial Engineer K. S. McHugh in "Bell System General Commercial Conference on Sales Matters, 1931," microfilm 368B, ILL BELL INFO. On the origins of in-house advertising, see N. L. Griese, "1908 Origins" (n. 11 above).

[17]See correspondence in "Advertising and Publicity—Bell System—1906–1910, Folder 1," box 1317, AT&T ARCH. Some reports claimed that thousands of stories were placed in hundreds of publications. Apparently no national advertising campaigns were conducted prior to these years; Bell marketing strategy seemed largely confined to price and service competition. See N. C. Kingsbury, "Results from the American Telephone's National Campaign," *Printers' Ink* (June 29, 1916): 182–84.

[18]In addition to the advertising collections, see A. P. Reynolds, "Selling a Telephone" (to a businessman), *Telephony* 12 (1906): 280–81; id., "The Telephone in Retail Business," *Printers' Ink* 61 (November 27, 1907): 3–8; and "Bell Encourages Shopping by Telephone," ibid., vol. 70 (January 19, 1910).

[19]Letter from AT&T Vice-President Reagan to PT&T President H. D. Pillsbury, March 4, 1929, in "Advertising," SF PION MU; W. J. Phillips, "The How, What, When and Why of Telephone Advertising," talk given July 7, 1926, in ibid.; and "Advertising Conference—Bell System—1916," box 1310, AT&T ARCH, p. 44.

shopping by telephone. Just one advertisement in this set explicitly suggests an amiable conversation: A grandmotherly woman is speaking on the telephone, a country vista visible through the window behind her, and says: "My! How sweet and clear my daughter's voice sounds! She seems to be right here with me!" The text reads: "Let us suggest a long distance visit home today." But this sort of advertisement was unusual.

During and immediately after World War I, there was no occasion to promote telephone use, since the industry struggled to meet demand pent up by wartime diversions. Much publicity tried to ease customer irritation at delays.

Only in the mid-1920s did AT&T and the Bell companies refocus their attention, for the first time in years, to sales efforts.[20] The system was a major advertiser, and Bell leaders actively discussed advertising during the 1920s. Copy focused on high-profit services, such as long distance and extension sets; modern "psychology," so to speak, influenced advertising themes; and Bell leaders became more sensitive to the competition from other consumer goods. Sociability suggestions increased, largely in the context of long-distance marketing.

In the United States, long-distance advertisements still overwhelmingly targeted business uses, but "visiting" with kin now appeared as a frequent suggestion. Bell Canada, for some reason, stressed family ties much more. Typical of the next two decades of Bell Canada's long-distance advertisements are these, both from 1921: "Why night calls are popular. How good it would sound to hear mother's voice tonight, he thought—for there were times when he was lonely—mighty lonely in the big city"; and "it's a weekly affair now, those fond intimate talks. Distance rolls away and for a few minutes every Thursday night the familiar voices tell the little family gossip that both are so eager to hear." Sales pointers to employees during this era often suggested providing customers with lists of their out-of-town contacts' telephone numbers.

In the 1920s, the advertising industry developed "atmosphere" techniques, focusing less on the product and more on its consequences for the consumer.[21] A similar shift may have begun in Bell's advertising, as well: "The Southwestern Bell Telephone Company has decided [in 1923] that it is selling something more vital than distance, speed or accuracy. . . . [T]he telephone . . . almost brings [people] face to face. It is the next best thing to personal contact. So the fundamental purpose of the current advertising is to sell the company's subscribers their voices at their true worth—to help them realize that 'Your Voice is You.'. . . to make subscribers think of the telephone whenever they think of distant friends or relatives. . . ."[22] This attitude was apparently only a harbinger, because during most of the 1920s the sociability theme was largely restricted to long distance and did not appear in many basic service advertisements.

Bell System salesmen spent the 1920s largely selling ancillary services, such as extension telephones, upgrading from party lines, and long distance, to current subscribers, rather than finding new customers. Basic residential rates averaged two to three dollars a month (about 2 percent of average manufacturing wages), not much different from a decade earlier, and Bell leaders did not consider seeking new subscribers to be sufficiently profitable to pursue seriously.[23] The limited new subscriber advertising continued the largely practical themes of earlier years. PT&T contended that residential telephones, especially extensions, were useful for emergencies, for social convenience (don't miss a call about an invitation, call your wife to set an extra place for dinner), and for avoiding the embarrassment of borrowing a telephone, as well as for its familiar business uses. A 1928 Bell Canada sales manual stressed household practicality first and social invitations second as tactics for selling basic service.[24]

---

[20]See n. 16 above.

[21]D. Pope, *The Making of Modern Advertising* (New York, 1983); S. Fox, *The Mirror Makers: A History of American Advertising and Its Creators* (New York, 1984); M. Schudson, *Advertising: The Uneasy Persuasion* (New York, 1985), pp. 60ff; R. Marchand, *Advertising the American Dream: Making Way for Modernity, 1920–1940* (Berkeley, Calif., 1985); and R. Pollay, "The Subsiding Sizzle: A Descriptive History of Print Advertising, 1900–1980," *Journal of Marketing* 49 (Summer 1985): 24–37.

[22]W. B. Edwards, "Tearing Down Old Copy Gods," *Printers' Ink* 123 (April 26, 1923): 65–66.

[23]On rates, see W. F. Gray, "Typical Schedules for Rates of Exchange Service," and related discussion, in "Bell System General Commercial Engineers' Conference, 1924," microfilm 364B, ILL BELL INFO.

[24]Bell Telephone Company of Canada, "Selling Service on the Job," ca. 1928, cat. 12223, BELL CAN HIST.

Then, in the late 1920s, Bell System leaders—prodded perhaps by the embarrassment that, for the first time, more American families owned automobiles, gas service, and electrical appliances than subscribed to telephones—pressed a more aggressive strategy. They built up a full-fledged sales force. And they sought to market the telephone as a "comfort and convenience"—that is, as more than a practical device—drawing somewhat on the psychological, sensualist themes in automobile advertising. They focused not only on upgrading the service of current subscribers but also on reaching those car owners and electricity users who lacked telephones. And the *social* character of the telephone was to be a key ingredient in the new sales strategies.[25]

Before "comfort and convenience" could go far, however, the Depression drew the industry's attention to basic service once again. Subscribers were disconnecting. Bell companies mounted campaigns to save residential connections by mobilizing *all* employees to sell or save telephone hookups on their own time (a program that had started before the Crash), expanding sales forces, advertising to current subscribers, and mounting door-to-door "save" and "nonuser" campaigns in some communities.[26] The "pitches" PT&T suggested to its employees included convenience (e.g., saving a trip to market), avoiding the humiliation of borrowing a neighbor's telephone, and simply being "modern." Salesmen actually seemed to rely more on pointing out the emergency uses of the telephone—an appeal especially telling to parents of young children—and suggesting that job offers might come via the telephone. Having a telephone so as to be available to friends and relatives was a lesser sales point. By now, a half-century since A. G. Bell's invention, salespeople did not have to sell telephone service itself but had to convince potential customers that they needed a telephone in their own homes.[27]

During the Depression, long-distance advertising continued, employing both business themes and the themes of family and friendship. But basic service advertising, addressed to both nonusers and would-be disconnectors, became much more common than it had been for twenty years.

The first line of argument in print ads for basic service was practicality—emergency uses, in particular—but suggestions for sociable conversations were more prominent than they had been before. A 1932 advertisement shows four people sitting around a woman who is speaking on the telephone. "Do Come Over!" the text reads, "Friends who are linked by telephone have good times." A 1934 Bell Canada advertisement features a couple who have just re-subscribed and who testify, "We got out of touch with all of our friends and missed the good times we have now." A 1935 advertisement asks, "Have you ever watched a person telephoning to a friend? Have you noticed how readily the lips part into smiles . . . ?" And 1939 copy states, "Some one thinks of some one, reaches for the telephone, and all is well." A 1937 AT&T advertisement reminds us that "the telephone is vital in emergencies, but that is not the whole of its service. . . . Friendship's path often follows the trail of the telephone wire." These family-and-friend motifs, more frequent and frank in the 1930s, forecast the jingles of today, such as ". . . a friendly voice, like chicken soup/is good for your health/Reach out, reach out and touch someone."[28]

[25]Comments, esp. by AT&T vice-presidents Page and Gherardi, during "General Commercial Conference, 1928," and "Bell System General Commercial Conference, 1930," both microfilm 368B, ILL BELL INFO, expressed a view that telephones should be part of consumers' "life-styles," not simply their practical instruments. One hears many echoes of "comfort and convenience" at lower Bell levels during this period.

[26]See A. Fancher, "Every Employee Is a Salesman for American Telephone and Telegraph," *Sales Management* 28 (February 26, 1931): 45–51, 472; "Bell Conferences," 1928 and 1930 (n. 25 above), esp. L. J. Billingsley, "Presention of Disconnections," in 1930 conference; *Pacemaker,* a sales magazine for PT&T, ca. 1928–31, SF PION MU; and *Telephony,* passim, 1931—36.

[27]PT&T *Pacemaker;* interviews by John Chan with retired industry executives in northern California; see also J. E. Harrel, "Residential Exchange Sales in New England Southern Area," in "Bell Conference, 1931" (n. 16 above), pp. 67ff.

[28]There is some variation among the advertising collections I examined. Illinois Bell's basic service advertisements used during the Depression are, for the most part, similar to basic service ads used a generation earlier. The Pacific Bell and Bell Canada advertisements feature sociable conversations much more. On the other hand, the Bell Canada ads are distinctive in that sociability is almost exclusively a family matter. Friendship, featured in U.S. ads all along, emerges clearly in the Canadian ads only in the 1930s. The 1932 ad cited in the text appears in the August 17 issue of the *Antioch* (Calif.) *Ledger.* The "chicken soup" jingle, sung by Roger Miller, was a Bell System ad in 1981. On the "Touch Someone" campaigns, see M. J. Arlen, *Thirty Seconds* (New York, 1980). See also "New Pitch to Spur Phone Use," *New York Times,* October 23, 1985, p. 44.

This brief chronology draws largely from prepared copy in industry archives, not from actual printed advertisements. A systematic survey, however, of two newspapers in northern California confirms the impression of increasing sociability themes. Aside from one 1911 advertisement referring to farm wives' isolation, the first sociability message in the *Antioch Ledger* appeared in 1929, addressed to parents: "No girl wants to be a wallflower." It was followed in the 1930s with notices for basic service such as "Give your friends straight access to your home," and "Call the folks now!" In 1911, advertisements in the *Marin* (County) *Journal* stressed the convenience of the telephone for automotive tourists. Sociability became prominent in both basic and long-distance advertisements in the late 1920s and the 1930s with suggestions that people "broaden the circle of friendly contact" (1927), "Voice visit with friends in nearby cities" (1930), and call grandmother (1935), and with the line, "I got my telephone for convenience. I never thought it would be such fun!" (1940).[29]

The emergence of sociability also appears in guides to telephone salesmen. A 1904 instruction booklet for sales representatives presents many selling points, but only one paragraph addresses residential service. That paragraph describes ways that the telephone saves time and labor, makes the household run smoothly, and rescues users in emergencies, but the only barely social use it notes is that the telephone "invites one's friends, asks them to stay away, asks them to hurry and enables them to invite in return." Conversation—telephone "visiting"—per se is not mentioned.

A 1931 memorandum to sales representatives, entitled "Your Telephone," is, on the other hand, full of tips on selling residential service and encouraging its use. Its first and longest subsection begins: *"Fosters friendships.* Your telephone will keep your personal friendships alive and active. Real friendships are too rare and valuable to be broken when you or your friends move out of town. Correspondence will help for a time, but friendships do not flourish for long on letters alone. When you can't visit in person, telephone periodically. Telephone calls will keep up the whole intimacy remarkably well. There is no need for newly-made friends to drop out of your life when they return to distant homes." A 1935 manual puts practicality and emergency uses first as sales arguments but explicitly discusses the telephone's "social importance," such as saving users from being "left high and dry by friends who can't reach [them] conveniently."[30]

This account, so far, covers the advertising of the Bell System. There is less known and perhaps less to know about the independent companies' advertising. Independents' appeals seem much like those of the Bell System, stressing business, emergencies, and practicality, except perhaps for showing an earlier sensitivity to sociability among their rural clientele.[31]

In sum, the variety of sales materials portray a similar shift. From the beginning to roughly the mid-1920s, the industry sold service as a practical business and household tool, with only occasional mention of social uses and those largely consisting of brief messages. Later sales arguments, for both long-distance and basic service, featured social uses prominently, including the suggestion that the telephone be used for conversations ("voice visiting") among friends and family. While it would be helpful to confirm this impressionistic

[29]These particular newspapers were examined as part of a larger study on the social history of the telephone that will include case studies of three northern California communities from 1890 to 1940.

[30]Central Union Telephone Company Contracts Department, *Instructions and Information for Solicitors,* 1904, ILL BELL INFO. Note that Central Union had been, at least through 1903, one of Bell's most aggressive solicitors of business. Illinois Bell Commercial Department, *Sales Manual* 1931, microfilm, ILL BELL INFO. Ohio Bell Telephone Company, "How You Can Sell Telephones," 1935, file "Salesmanship," BELL CAN HIST.

[31]Until 1894, independent companies did not exist. For years afterward, they largely tried to meet unfilled demand in the small cities and towns Bell had underserved. In other places, they advertised competitively against Bell. Nevertheless, advertising men often exhorted the independents to use "salesmanship in print" to encourage basic service and extensive use. See, e.g., J. A. Schoell, "Advertising and Other Thoughts of the Small Town Man," *Telephony* 70 (June 10, 1916): 40–41; R. D. Mock, "Fundamental Principles of the Telephone Business: Part V, Telephone Advertising," series in ibid., vol. 71 (July 22–November 21, 1916); D. Hughes, "Right Now Is the Time to Sell Service," ibid., 104 (June 10, 1933): 14–15; and L. M. Berry, "Helpful Hints for Selling Service," ibid., 108 (February 2, 1935): 7–10. See also Kellogg Company, "A New Business Campaign for _____" (Chicago: Kellogg, 1929), MU IND TEL.

account with firm statistics, for various reasons it is difficult to draw an accurate sample of advertising copy and salesmen's pitches for over sixty years. (For one, we have no easily defined "universe" of advertisements. Are the appropriate units specific printed ads, or ad campaigns? How are duplicates to be handled? Or ads in neighboring towns? Do they include planted stories, inserts in telephone bills, billboards, and the like? Should locally generated ads be included? And what of nationally prepared ads not used by the locals? For another, we have no clear "population" of ads. The available collections are fragmentary, often preselected for various reasons.) An effort in that direction appears, however, in table 2, in which the numbers of "social" advertisements show a clear increase, both absolutely and relatively.

TABLE 2

COUNTS OF DOMINANT ADVERTISING THEMES BY PERIOD

Sources and Types of Advertisements	Prewar		1919–29		1930–40	
Antioch (Calif.) Ledger:						
Social, sociability	1	(1)	1	(1)	6	(4)
Business, businessmen	6	(5)	1	(1)	2	(1)
Household, convenience, etc.	5	(5)	3	(3)	4	(3)
Public relations, other	0	(0)	4	(3)	1	(1)
Total	12	(11)	9	(8)	13	(9)
Approximate ratio of social to others	1:11	(1:10)	1:8	(1:7)	1:1	(1:1)
Marin (Calif.) Journal:						
Social, sociability	1	(1)	5	(2)	43	(20)
Business, businessmen	2	(2)	8	(2)	10	(3)
Household, convenience, etc.	12	(12)	3	(3)	20	(20)
Public relations, other	0	(0)	19	(13)	25	(16)
Total	15	(15)	35	(20)	98	(59)
Approximate ratio of social to others	1:14	(1:14)	1:6	(1:9)	1:1	(1:2)
Bell Canada:						
Social, sociability	5	(2)	25	(1)	59*	(9)
Business, businessmen	20*	(20)	15	(2)	24*	(4)
Household, convenience, etc.	28	(28)	3	(3)	23*	(6)
Public relations, other	30*	(30)	25	(40)	2	(2)
Total	83*	(80)	68	(46)	108*	(21)
Approximate ratio of social to others	1:16	(1:39)	1:2	(1:45)	1:1	(1:1)
Pacific Telephone, 1914–15:						
Social, sociability	2	(1)	. . .		. . .	
Business, businessmen	7	(6)	. . .		. . .	
Household, convenience, etc.	18	(16)	. . .		. . .	
Public relations, other	16	(9)	. . .		. . .	
Total	43	(32)	. . .		. . .	
Approximate ratio of social to others	1:21	(1:31)	. . .		. . .	

TABLE 2 (*continued*)

Sources and Types of Advertisements	Prewar		1919–29	1930–40
Assorted Bell ads, 1906–10:				
Social, sociability.................................	4	(4)	. . .	. . .
Business, businessmen.........................	13	(12)	. . .	. . .
Household, convenience, etc.	11	(11)	. . .	. . .
Public relations, other..........................	9	(9)	. . .	. . .
Total	37	(36)	. . .	. . .
Approximate ratio of social to others	1:8	(1:8)	. . .	. . .

SOURCES—Advertisements in the *Antioch Ledger* were sampled from 1906 to 1940 by Barbara Loomis; those in the *Marin Journal* were sampled from 1900 to 1940 by John Chan. The Bell Canada collection appears in scrapbooks at Bell Canada Historical; the Pacific collection is in the San Francisco Pioneer Telephone Museum. The AT&T advertisements are from AT&T ARCH, box 1317. Other, spotty collections were used for the study but not counted here because they were not as systematic. All coding was done by the author.

NOTE—Counts in parentheses exclude explicitly long-distance advertisements. Usually each ad had one dominant theme. When more than one seemed equal in weight, the ad was counted in both categories. "Social, sociability" refers to the use of the telephone for personal contact, including season's greetings, invitations, and conversation between friends and family. (Note that the inclusion of brief messages in this category makes the analysis a conservative test of the argument that there was a shift toward sociability themes.) "Business, businessmen" refers to the explicit use of the telephone for business purposes or general appeals to businessmen—e.g., that the telephone will make one a more forceful entrepreneur. "Household, convenience, etc." includes the use of the telephone for household management, personal convenience (e.g., don't get wet, order play tickets), and for emergencies, such as illness or burglary. "Public relations, other" includes general institutional advertising, informational notices (such as how to use the telephone), and other miscellaneous. Perhaps the most conservative index is the ratio of non-long-distance social ads to non-long-distance household ads. (Business ads move to speciality magazines over the years; public information ads fluctuate with political events; and long-distance ads may be "inherently" social.) In the *Antioch Ledger,* this ratio changes from 1:5 to 4:3; in the *Marin Journal,* from 1:12 to 1:1; and in Bell Canada's ads, from 1:14 to 1.5:1. Even these ratios understate the shift, for several reasons. One, I was much more alert to social than to other ads and was more thorough with early social ads than any other category. Two, the household category is increased in the later years by numerous ads for extension telephones. Three, the nature of the social ads counted here changes. The earlier ones overwhelmingly suggest using the telephone for greetings and invitations, not conversation. With rare exception, only the later ones discuss friendliness and "warm human relationships" and suggest chats.

* Estimated.

▪ Industry Attitudes toward Sociability

This change in advertising themes apparently reflected a change in the actual beliefs industry men held about the telephone. Alexander Graham Bell himself forecast social chitchats using his invention. He predicted that eventually Mrs. Smith would spend an hour on the telephone with Mrs. Brown "very enjoyably . . . cutting up Mrs. Robinson."[32] But for decades few of his successors saw it that way.

Instead, the early telephone vendors often battled their residential customers over social conversations, labeling such calls "frivolous" and "unnecessary." For example, an 1881 announcement complained, "The fact that subscribers have been free to use the wires as they pleased without incurring additional expense [i.e., flat rates] has led to the transmission of large numbers of communications of the most trivial character."[33] In 1909, a local telephone manager in Seattle listened in on a sample of conversations coming through a residential exchange and determined that 20 percent of the calls were orders to stores and other businesses, 20 percent were from subscribers' homes to their own businesses, 15 percent were social invitations, and 30 percent were "purely idle gossip"—a rate that he claimed was matched in other cities. The manager's concern was to reduce this last, "unnecessary use." One tactic for doing so, in addition to "education" campaigns on proper use of the telephone, was to place time limits on calls (in his survey the average call had lasted over seven minutes). Time limits were often an explicit effort to stop people who insisted on chatting when there was "business" to be conducted.[34]

---

[32]Quoted in Aronson, "Electrical Toy" (n. 14 above).

[33]Proposed announcement by National Capitol Telephone Company, in letter to Bell headquarters, January 20, 1881, box 1213, AT&T ARCH. In a similar vein, the president of Bell Canada confessed, ca. 1890, to being unable to stop "trivial conversations"; see Collins, *A Voice* (n. 2 above), p. 124. The French authorities were also exasperated by nonserious uses; see C. Bertho, *Télégraphes et téléphones* (Paris, 1980), pp. 244–45.

[34]C. H. Judson, "Unprofitable Traffic—What Shall Be Done with It?" *Telephony* 18 (December 11, 1909): 644–47, and PAC TEL MAG 3 (January, 1910): 7. He also writes, "the telephone is going beyond its original design, and it is a positive fact that a large

An exceptional few in the industry, believing in a more "populist" telephony, did, however, try to encourage such uses. E. J. Hall, Yale-educated and originally manager of his family's firebrick business, initiated the first "measured service" in Buffalo in 1880 and later became an AT&T vice-president. A pleader for lower rates, Hall also defended "trivial" calls, arguing that they added to the total use-value of the system. But the evident isolation of men like Hall underlines the dominant antisociability view of the pre-World War I era.[35]

Official AT&T opinions came closer to Hall's in the later 1920s when executives announced that, whereas the industry had previously thought of telephone service as a practical necessity, they now realized that it was more: it was a "convenience, comfort, luxury"; its value included its "trivial" social uses. In 1928, Publicity Vice-President A. W. Page, who had entered AT&T from the publishing industry the year before, was most explicit when he criticized earlier views: "There had also been the point of view [in the Bell System and among the public] about not using the telephone for frivolous conversation. This is about as commercial as if the automobile people should advertise. 'Please do not take out this car unless you are going on a serious errand. . . .' We are faced, I think, with a state of public consciousness that the telephone is a necessity and not to be trifled with, certainly in the home." Bell sales officials were told to sell telephone service as a "comfort and convenience," including as a conversational tool.[36]

Although this change in opinion is most visible for the Bell System, similar trends can be seen in the pages of the journal of the independent companies, *Telephony*, especially in regard to rural customers. Indeed, early conflict about telephone sociability was most acute in rural areas. During the monopoly era, Bell companies largely neglected rural demand. The depth and breadth of that demand became evident in the first two decades of this century, when proportionally more farm than urban households obtained telephones, the former largely from small commercial or cooperative local companies. Sociability both spurred telephone subscription and irritated the largely non-Bell vendors.

The 1907 Census of Telephones argued that in areas of isolated farmhouses "a sense of community life is impossible without this ready means of communication. . . . The sense of loneliness and insecurity felt by farmers' wives under former conditions disappears, and an approach is made toward the solidarity of a small country town." Other official investigations bore similar witness.[37] Rural telephone men also dwelt on sociability. One independent company official stated: "When we started the farmers thought they could get along without telephones. . . . Now you couldn't take them out. The women wouldn't let you even if the men would. Socially, they have been a godsend. The women of the county keep in touch with each other, and with their social duties, which are largely in the nature of church work."[38]

---

percentage of telephones in use today on a flat rental basis are used more in entertainment, diversion, social intercourse and accommodation to others, than in actual cases of business or household necessity" (p. 645). MacMeal, *Independent* (n. 2 above), p. 240, reports on a successful campaign in 1922 to discourage gossipers through letters and advertisements. Typically, calls were—at least officially—limited to five minutes in many places, although it is unclear how well limits were enforced.

[35]Hall's philosophy is evident in the correspondence over measured service before 1900, box 1127, AT&T ARCH. Decades later, he pushed it in a letter to E. M. Burgess, Colorado Telephone Company, March 30, 1905, box 1309, AT&T ARCH, even arguing that operators should stop turning away calls made by children and should instead encourage such "trivial uses." The biographical information comes from an obituary in AT&T ARCH. Another, more extreme populist was John L. Sabin, of PT&T and the Chicago Telephone Co.; see Mahon (n. 2 above), pp. 29ff.

[36]A. W. Page, "Public Relations and Sales," "General Commercial Conference, 1928," p. 5, microfilm 368B, ILL BELL INFO. See also comments by Vice-President Gherardi and others in same conference and related ones of the period. On Page and the changes he instituted, see G. J. Griswold, "How AT&T Public Relations Policies Developed," *Public Relations Quarterly* 12 (Fall 1967): 7–16; and Marchand, *Advertising* (n. 21 above), pp. 117–20.

[37]BOC, *Special Reports: Telephones: 1907* (Washington, D.C., 1910), pp. 77–78; see also U.S. Congress, Senate, Country Life Commission, 60th Cong., 2d sess., 1909, S. Doc. 705; and F. E. Ward, *The Farm Woman's Problems*, USDA Circular 148 (Washington, D.C., 1920). See also C. S. Fischer, "The Revolution in Rural Telephony," *Journal of Social History* (in press).

[38]Quoted in R. F. Kemp, "Telephones in Country Homes," *Telephony* 9 (June 1905): 433. A 1909 article claims that "[t]he principle use of farm line telephones has been their social use. . . . The telephones are more often and for longer times held for neighborly conversations than for any other purpose." It goes on to stress that subscribers valued conversation with anyone on the line; see G. R. Johnston, "Some Aspects of Rural Telephony," *Telephony* 17 (May 8, 1909): 542. See also R. L. Tomblen, "Recent Changes in Agriculture as Revealed by the Census," *Bell Telephone Quarterly* 9 (October 1932): 334–50; and J. West (C. Withers), *Plainville, U.S.A.* (New York, 1945), p. 10.

Although the episodic sales campaigns to farmers stressed the practical advantages of the telephone, such as receiving market prices, weather reports, and emergency aid, the industry addressed the social theme more often to them than to the general public. A PT&T series in 1911, for example, focused on the telephone in emergencies, staying informed, and saving money. But one additional advertisement said it was: "A Blessing to the Farmer's Wife. . . . It relieves the monotony of life. She CANNOT be lonesome with the Bell Service. . . ."[39] For all that, telephone professionals who dealt with farmers often fought the use of the line for nonbusiness conversations, at least in the early years. The pages of *Telephony* overflow with complaints about farmers on many grounds, not the least that they tied up the lines for chats.

More explicit appreciation of the value of telephone sociability to farmers emerged later. A 1931 account of Bell's rural advertising activities stressed business uses, but noted that "only within recent years [has] emphasis been given to [the telephone's] usefulness in everyday activities . . . the commonplaces of rural life." A 1932 article in the *Bell Telephone Quarterly* notes that "telephone usage for social purposes in rural areas is fundamentally important." Ironically, in 1938, an independent telephone man claimed that the social theme *had been* but was *no longer* an effective sales point because the automobile and other technologies had already reduced farmers' isolation![40]

As some passages suggest, the issue of sociability was also tied up with gender. When telephone vendors before World War I addressed women's needs for the telephone, they usually meant household management, security, and emergencies. There is evidence, however, that urban, as well as rural, women found the telephone to be useful for sociability.[41] When industry men criticized chatting on the telephone, they almost always referred to the speaker as "she." Later, in the 1930s, the explicit appeals to sociability also emphasized women; the figures in such advertisements, for example, were overwhelmingly women.

In rough parallel with the shift in manifest advertising appeals toward sociability, there was a shift in industry attitudes from irritation with to approval of sociable conversations as part of the telephone's "comfort, convenience, and luxury."

■ Economic Explanations

Why were the telephone companies late and reluctant to suggest sociable conversations as a use? There are several, not mutually exclusive, possible answers. The clearest is that there was no profit in sociability at first but profit in it later.

Telephone companies, especially Bell, argued that residential service had been a marginal or losing proposition, as measured by the revenues and expenses accounted to each instrument, and that business service had subsidized local residential service. Whether this argument is valid remains a matter of debate. Nevertheless, the belief that residential customers were unprofitable was common, especially among line workers, and no doubt

[39]The PT&T series appeared in the *Antioch* (Calif.) *Ledger* in 1911. For some examples and discussions of sales strategies to farmers, see Western Electric, "How to Build Rural Lines," n.d., "Rural Telephone Service, 1944–46," box 1310, AT&T ARCH; Stromberg-Carlson Telephone Manufacturing Company, *Telephone Facts for Farmers* (Rochester, N.Y., 1903), Warshaw Collection, Smithsonian Institution; "Facts regarding the Rural Telephone," *Telephony* 9 (April 1905): 303. In *Printers' Ink,* "The Western Electric," 65 (December 23, 1908): 3–7; F. X. Cleary, "Selling to the Rural District," 70 (February 23, 1910): 11–12; "Western Electric Getting Farmers to Install Phones," 76 (July 27, 1911): 20–25; and H. C. Slemin, "Papers to Meet 'Trust' Competition," 78 (January 18, 1912): 28.

[40]R. T. Barrett, "Selling Telephones to Farmers by Talking about Tomatoes," *Printers' Ink* (November 5, 1931): 49–50; Tomblen (n. 38 above); and J. D. Holland, "Telephone Service Essential to Progressive Farm Home," *Telephony* 114 (February 19, 1938): 17–20. See also C. S. Fischer, "Technology's Retreat: The Decline of Rural Telephones, 1920–1940," *Social Science History* (in press).

[41]A 1925 survey of women's attitudes toward home appliances by the General Federation of Women's Clubs showed that respondents preferred automobiles and telephones above indoor plumbing; see M. Sherman, "What Women Want in Their Homes," *Woman's Home Companion* 52 (November 1925): 28, 97–98. A census survey of 500,000 homes in the mid-1920s reportedly found that the telephone was considered a primary household appliance because it, with the automobile and radio, "offer[s] the homemaker the escape from monotony which drove many of her predecessors insane"; reported in *Voice Telephone Magazine,* in-house organ of United Communications, December 1925, p. 3, MU IND TEL. One of our interviewees who conducted door-to-door telephone sales in the 1930s said that women were attracted to the service first in order to talk to kin and friends, second for appointments and shopping, and third for emergencies, while, for men, employment and business reasons ranked first. See also Rakow, "Gender" (n. 2 above), and C. S. Fischer, "Women and the Telephone, 1890–1940," paper presented to the American Sociological Association, 1987.

discouraged intensive sales efforts to householders.[42] At times, Bell lacked the capital to construct lines needed to meet residential demand. These constraints seemed to motivate occasional orders from New York not to advertise basic service or to do so only to people near existing and unsaturated lines.[43] And, at times, there was a technical incompatibility between the quality of service Bell had accustomed its business subscribers to expect and the quality residential customers were willing to pay for. Given these considerations, Bell preferred to focus on the business class, who paid higher rates, bought additional equipment, and made long-distance calls.[44]

Still, when they did address residential customers, why did telephone vendors not employ the sociability theme until the 1920s, relying for so long only on practical uses? Perhaps social calls were an untouched and elastic market of consumer demand. Having sold the service to those who might respond to practical appeals— and perhaps by World War I everyone knew those practical uses—vendors might have thought that further expansion depended on selling "new" social uses of the telephone.[45] Similarly, vendors may have thought they had already enrolled all the subscribers they could—42 percent of American households in 1930—and shifted attention to encouraging use, especially of toll lines. We have seen how sales efforts for intercity calls invoked friends and family. But this explanation does not suffice. It leaves as a puzzle why the sociability themes continued in the Depression when the industry focused again on simply ensuring subscribers and also why the industry's internal attitudes shifted as well.

Perhaps the answer is in the rate structures. Initially, telephone companies charged a flat rate for unlimited local use of the service. In such a system, extra calls and lengthy calls cost users nothing but are unprofitable to providers because they take operator time and, by occupying lines, antagonize other would-be callers. Some industry men explicitly blamed "trivial" calls on flat rates.[46] Discouraging "visiting" on the telephone then made sense.

Although flat-rate charges continued in many telephone exchanges, especially smaller ones, throughout the period, Bell and others instituted "measured service" in full or in part—charging additionally per call—in most large places during the era of competition. In St. Louis in 1898, for example, a four-party telephone cost forty-five dollars a year for 600 calls a year, plus eight cents a call in excess.[47] This system allowed companies to reduce basic subscription fees and thus attract customers who wanted the service only for occasional use.

Company officials had conflicting motives for pressing measured service. Some saw it simply as economically rational, charging according to use. Others saw it as a means of reducing "trivial" calls and the borrowing of telephones by nonsubscribers. A few others, such as E. J. Hall, saw it as a vehicle for bringing in masses of small users.

The industry might have welcomed social conversations, if it could charge enough to make up for uncompleted calls and for the frustrated subscribers busy lines produced. In principle, under measured service, it could.

[42]See, e.g., J. W. Sichter, "Separations Procedures in the Telephone Industry," paper P-77-2, Harvard University Program on Information Resources (Cambridge, Mass., 1977); *Public Utilities Digest,* 1930s–1940s, passim; "Will Your Phone Rates Double?" *Consumer Reports* (March 1984): 154–56. Chan's industry interviewees believed this cross subsidy to be true, as, apparently, did AT&T's commercial engineers; see various "Conferences" cited above, AT&T ARCH and ILL BELL INFO.

[43]E.g., commercial engineer C. P. Morrill wrote in 1914 that "we are not actively seeking new subscribers except in a few places where active competition makes this necessary. Active selling is impossible due to rapid growth on the Pacific Coast." He encouraged sales of party lines in congested areas, individual lines in place of party lines elsewhere, extensions, more calling, directory advertisements, etc., rather than expanding basic service into new territories; see PAC TEL MAG 7 (1914): 13–16. And, in 1924, the Bell System's commercial managers decided to avoid canvassing in areas that would require plant expansion and to stress instead long-distance calls and services, especially for large business users; see correspondence from B. Gherardi, vice-president, AT&T, to G. E. McFarland president, PT&T, July 14, 1924, and November 26, 1924, folder "282—Conferences," SF PION MU, and exchange with McFarland, May 10 and May 20, 1924, folder "Correspondence—B. Gherardi," SF PION MU.

[44]The story of the Chicago exchange under John L. Sabin illustrates the point. See R. Garnet, "The Central Union Telephone Company," box 1080, AT&T ARCH.

[45]This point was suggested by John Chan from the interviews.

[46]See n. 33, 34. This is also the logic of a recent New York Telephone Co. campaign to encourage social calls: The advertising will not run in upstate New York "since the upstaters tend to have flat rates and there would be no profit in having them make unnecessary calls" (see "New Pitch," n. 28 above).

[47]Letter to AT&T President Hudson, December 27, 1898, box 1284, AT&T ARCH. On measured service in general, see "Measured Service Rates," boxes 1127, 1213, 1287, 1309, AT&T ARCH; F. H. Bethell, "The Message Rate," repr. 1913, AT&T ARCH; H. B. Stroud, "Measured Telephone Service," *Telephony* 6 (September 1903): 153–56, and (October 1903): 236–38; and J. E. Kingsbury, *The Telephone and Telephone Exchanges* (London, 1915), pp. 469–80.

(As it could with long distance, where each minute was charged.) Although mechanical time metering was apparently not available for most or all of this period, rough time charges for local calls existed in principle, since "messages" were typically defined as five minutes long or any fraction thereof. Thus, "visiting" for twenty minutes should have cost callers four "messages." In such systems, the companies would have earned income from sociability and might have encouraged it.[48]

However, changes from flat rates to measured rates do *not* seem to explain the shift toward sociability around the 1920s. Determining the extent that measured service was actually used for urban residential customers is difficult because rate schedules varied widely from town to town even within the same states. But the timing does not fit. The big exchanges with measured residential rates had them early on. For example, in 1904, 96 percent of Denver's residential subscribers were on at least a partial measured system, and, in 1905, 90 percent of those in Brooklyn, New York, were as well. (Yet, Los Angeles residential customers continued to have flat rates.)[49] There is little sign that these rate systems altered significantly in the next twenty-five years while sociability themes emerged.

Conversely, flat rates persisted in small exchanges beyond the 1930s. Moreover, sociability themes appeared more often in rural sales campaigns than in urban ones, despite the fact the rural areas remained on flat-rate schedules.

Although concern that long social calls occupied lines and operators—with financial losses to the companies—no doubt contributed to the industry's resistance to sociability, it is not a sufficient explanation of those attitudes or, especially, of the timing of their change.

■ Technical Explanations

Industry spokesmen early in the era would probably have claimed that technical considerations limited "visiting" by telephone. Extended conversations monopolized party lines. That is why companies, often claiming customer pressure, encouraged, set—or sought legal permission to set—time limits on calls. Yet, this would not explain the shift toward explicit sociability, because as late as 1930, 40–50 percent of Bell's main telephones in almost all major cities were still on party lines, a proportion not much changed from 1915.[50]

A related problem was the tying up of toll lines among exchanges, especially those among villages and small towns. Rural cooperatives complained that the commercial companies provided them with only single lines between towns. The companies resisted setting up more, claiming they were underpaid for that service. This single-line connection would create an incentive to suppress social conversations, at least in rural areas. But this does not explain the shift toward sociability either. The bottleneck was resolved much later than the sales shift when it became possible to have several calls on a single line.[51]

---

[48]Theodore Vail claimed in 1909 that mechanical time metering was impossible (in testimony to a New York State commission, see n. 8 above, p. 470). See also Judson (n. 34 above), p. 647. In 1928, an operating engineer suggested overtime charges on five-minute calls and stated that equipment for monitoring overtime was now available; see L. B. Wilson, "Report on Commercial Operations, 1927," in "General Commercial Conference, 1928," p. 28, microfilm 368B, ILL BELL INFO. On the five-minute limit, see "Measured Service," box 1127, AT&T ARCH, passim; and Bell Canada, *The First Century of Service* (Montreal, 1980), p. 4. There is no confirmation on how strict operators in fact were in charging overtime. The Bell System, at least, was never known for its laxness in such matters.

[49]Denver: letter from E. J. Hall to E. W. Burgess, 1905, box 1309, AT&T ARCH; Brooklyn: BOC, *Telephones, 1902* (n. 4 above); Los Angeles: "Telephone on the Pacific Coast, 1878–1923," box 1045, AT&T ARCH.

[50]On company claims, see, e.g., "Limiting Party Line Conversations," *Telephony* 66 (May 2, 1914): 21; and MacMeal (n. 2 above), p. 224. On party-line data, compare the statistics in the letter from J. P. Davis to A. Cochrane, April 2, 1901, box 1312, AT&T ARCH, to those in B. Gherardi and F. B. Jewett, "Telephone Communications System of the United States," *Bell System Technical Journal* 1 (January 1930): 1–100. The former show, e.g., that, in 1901, in the five cities with the most subscribers, an average of 31 percent of telephones were on party lines. For those five cities in 1929, the percentage was 36. Smaller exchanges tended to have even higher proportions. See also "Supplemental Telephone Statistics, PT&T," "Correspondence—Du Bois," SF PION MU. The case of Bell Canada also fails to support a party-line explanation. Virtually all telephones in Montreal and Toronto were on individual lines until 1920.

[51]"Carrier currents" allowed multiple conversations on the same line. The first one was developed in 1918, but for many years they were limited to use on long-distance trunk lines, not local toll lines. See, e.g., R. Coe, "Some Distinguishing Characteristics of the Telephone Business," *Bell Telephone Quarterly* 6 (January 1927): 47—51, esp. pp. 49–50; and R. C. Boyd, J. D. Howard, Jr., and L. Pederson, "A New Carrier System for Rural Service," *Bell System Technical Journal* 26 (March 1957): 349–90. The first long-distance carrier line was established in Canada in 1928, after the long-distance sociability theme had emerged; see Bell Canada, *First Century,* no. 46, p. 28.

The development of long distance might also explain increased sociability selling. Over the period covered here, the technology improved rapidly, AT&T's long-distance charges dropped, and its costs dropped even more. The major motive for residential subscribers to use long distance was to greet kin or friends. Additionally, overtime was well monitored and charged. Again, while probably contributing to the overall frequency of the sociability theme, long-distance development seems insufficient to explain the change. Toll calls as a proportion of all calls increased from 2.5 percent in 1900 to 3.2 percent in 1920 and 4.1 percent in 1930, then dropped to 3.3 percent in 1940. They did not reach even 5 percent of all calls until the 1960s.[52] More important, the shift toward sociability appears in campaigns to sell basic service and to encourage local use, as well as in long-distance ads. (See table 2.)

▪ Cultural Explanations

While both economic and technical considerations no doubt framed the industry's attitude toward sociability, neither seems sufficient to explain the historical change. Part of the explanation probably lies in the cultural "mind-set" of the telephone men.

In many ways, the telephone industry descended directly from the telegraph industry. The instruments are functionally very similar; technical developments sometimes applied to both. The people who developed, built, and marketed telephone systems were predominantly telegraph men. Theodore Vail himself came from a family involved in telegraphy and started his career as a telegrapher. (In contrast, E. J. Hall and A. W. Page, among the supporters of "triviality," had no connections to telegraphy. J. L. Sabin, a man of the same bent, did have roots in telegraphy.) Many telephone companies had started as telegraph operations. Indeed, in 1880, Western Union almost displaced Bell as the telephone company. And the organization of Western Union served in some ways as a model for Bell. Telephone use often directly substituted for telegraph use. Even the language used to talk about the telephone revealed its ancestry. For example, an early advertisement claimed that the telephone system was the "cheapest telegraph service ever." Telephone calls were long referred to as "messages." American telegraphy, finally, was rarely used even for brief social messages.[53]

No wonder, then, that the uses proposed first and for decades to follow largely replicated those of a printing telegraph: business communiqués, orders, alarms, and calls for services. In this context, industry men reasonably considered telephone "visiting" to be an abuse or trivialization of the service. Internal documents suggest that most telephone leaders typically saw the technology as a business instrument and a convenience for the middle class, claimed that people had to be sold vigorously on these marginal advantages, and believed that people had no "natural" need for the telephone—indeed, that most (the rural and working class) would never need it. Customers would have to be "educated" to it.[54] AT&T Vice-President Page was reacting precisely against this telegraphy perspective in his 1928 defense of "frivolous" conversation. At the same conference, he also decried the psychological effect of telephone advertisements that explicitly compared the instrument to the telegraph.[55]

[52]BOC, *Historical Statistics* (n. 3 above), p. 783.

[53]On the telegraph background of early telephone leaders, see, e.g., A. B. Paine, *Theodore N. Vail* (New York, 1929); Rippey (n. 2 above); and W. Patten, *Pioneering the Telephone in Canada* (Montreal: Telephone Pioneers, 1926). Interestingly, this was true of Bell and the major operations. But the leaders of small-town companies were typically businessmen and farmers; see, e.g., *On the Line* (Madison: Wisconsin State Telephone Association, 1985). On Western Union and Bell, see G. D. Smith, *The Anatomy of a Business Strategy: Bell, Western Electric, and the Origins of the American Telephone Industry* (Baltimore, 1985). The "cheapest telegraph" appears in a Buffalo flier of November 13, 1880, box 1127, AT&T ARCH. On the infrequent use of the telegraph for social messages, see R. B. DuBoff, "Business Demand and Development of the Telegraph in the United States, 1844–1860," *Business History Review* 54 (Winter 1980): 459–79.

[54]In the very earliest days, Vail had expected that the highest level of development would be one telephone per 100 people; by 1880, development had reached four per 100 in some places; see Garnet (n. 2 above), p. 133, n. 3. It reached one per 100 Americans before 1900 (see table 1). In 1905, a Bell estimate assumed that twenty telephones per 100 Americans was the saturation point and even that "may appear beyond reason"; see "Estimated Telephone Development, 1905–1920," letter from S. H. Mildram, AT&T, to W. S. Allen, AT&T, May 22, 1905, box 1364, AT&T ARCH. The saturation date was forecast for 1920. This estimate was optimistic in its projected *rate* of diffusion—twenty per 100 was reached only in 1945—but very pessimistic in its projected *level* of diffusion. That level was doubled by 1960 and tripled by 1980. One reads in Bell documents of the late 1920s of concern that the automobile and other new technologies were far outstripping telephone diffusion. Yet, even then, there seemed to be no assumption that the telephone would reach the near universality in American homes of, say, electricity or the radio.

[55]Page 53 in L. B. Wilson (chair), "Promoting Greater Toll Service," "General Commercial Conference, 1928," microfilm 368B, ILL BELL INFO.

Industry leaders long ignored or repressed telephone sociability—for the most part, I suggest, because such conversations did not fit their understandings of what the technology was supposed to be for. Only after decades of customer insistence on making such calls—and perhaps prodded by the popularity of competing technologies, such as the automobile and radio—did the industry come to adopt sociability as a means of exploiting the technology.

This argument posits a generation-long lag, a mismatch, between how subscribers used the telephone and how industry men thought it would be used. A variant of the argument (posed by several auditors of this article) suggests that there was no mismatch, that the industry's attitudes and advertising accurately reflected public practice. Sales strategies changed toward sociability around the mid-1920s because, in fact, people began using the telephone that way more. This increase in telephone visiting occurred for perhaps one or more reasons—a drop in real costs, an increase in the number of subscribers available to call, clearer voice transmission, more comfortable instruments (from wall sets to the "French" handsets), measured rates, increased privacy with the coming of automatic dial switching, and so on—and the industry's marketing followed usage.

To address this argument fully would require detailed evidence on the use of the telephone over time, which we do not have. Recollections by some elderly people suggest that they visited by telephone less often and more quickly in the "old days," but they cannot specify exact rates or in what era practices changed.[56] On the other hand, anecdotes, comments by contemporaries, and fragments of numerical data (e.g., the 1909 Seattle "study") suggest that residential users regularly visited by telephone before the mid-1920s, whatever the etiquette was supposed to be, and that such calls at least equaled calls regarding household management. Yet, telephone advertising in the period overwhelmingly stressed practical use and ignored or suppressed sociability use.

Changes in customers' practices may have helped spur a change in advertising—although there is no direct evidence of this in the industry archives—but some sort of mismatch existed for a long time between actual use and marketing. Its source appears to be, in large measure, cultural.

This explanation gains additional plausibility from the parallel case of the automobile, about which space permits only brief mention. The early producers of automobiles were commonly former bicycle manufacturers who learned their production techniques and marketing strategies (e.g., the dealership system, annual models) during the bicycle craze of the 1890s. As the bicycle was then, so was the automobile initially a plaything of the wealthy. The early sales campaigns touted the automobile as a leisure device for touring, joyriding, and racing. One advertising man wondered as late as 1906 whether "the automobile is to prove a fad like the bicycle or a lasting factor in the industry of the country."[57]

That the automobile had practical uses dawned on the industry quickly. Especially after the success of the Ford Model T, advertisements began stressing themes such as utility and sociability—in particular, that families could be strengthened by touring together. Publicists and independent observers alike praised the automobile's role in breaking isolation and increasing community life.[58] As with the telephone, automobile vendors largely followed a marketing strategy based on the experience of their "parent" technology; they

[56]This comment is based on the oral histories reported by Rakow (n. 2 above) and by several interviews conducted in San Rafael, Calif., by John Chan for this project. See also Fischer, "Women" (n. 41).

[57]Among the basic sources on the history of the automobile drawn from are: J. B. Rae, *The American Automobile: A Brief History* (Chicago, 1965); id., *The Road and Car in American Life* (Cambridge, Mass., 1971); J. J. Flink, *America Adopts the Automobile, 1895–1910* (Cambridge, Mass., 1970); id., *The Car Culture* (Cambridge, Mass., 1976); and J.-P. Bardou, J.-J. Chanaron, P. Fridenson, and J. M. Laux, *The Automobile Revolution*, trans. J. M. Laux (Chapel Hill, N.C., 1982). The advertising man was J. H. Newmark, "Have Automobiles Been Wrongly Advertised?" *Printers' Ink* 86 (February 5, 1914): 70–72. See also id., "The Line of Progress in Automobile Advertising," ibid., 105 (December 26, 1918): 97–102.

[58]G. L. Sullivan, "Forces That Are Reshaping a Big Market," *Printers' Ink* 92 (July 29, 1915): 26–28. Newmark (n. 57 above, p. 97) wrote in 1918 that it "has taken a quarter century for manufacturers to discover that they are making a utility." A 1930s study suggested that 80 percent of household automobile expenditures was for "family living"; see D. Monroe et al., *Family Income and Expenditures. Five Regions*, Part 2. *Family Expenditures*, Consumer Purchases Study, Farm Series, Bureau of Home Economics, Misc. Pub. 465 (Washington, D.C., 1941), pp. 34–36. Recall the 1925 survey of women's attitudes toward appliances (n. 41 above). The author of the report, Federation President Mary Sherman, concluded that "Before toilets are installed or washbasins put into homes, automobiles are purchased and telephones are connected . . . [b]ecause the housewife for generations has sought escape from the monotony rather than the drudgery of her lot" (p. 98). See also *Country Life* and Ward (n. 37 above); E. de S. Brunner and J. H. Kolb, *Rural Social Trends* (New York, 1933); and F. R. Allen, "The Automobile," pp. 107–32 in F. R. Allen et al., *Technology and Social Change* (New York, 1957).

stressed a limited and familiar set of uses; and they had to be awakened, it seems, to wider and more popular uses. The automobile producers learned faster.

No doubt other social changes also contributed to what I have called the discovery of sociability, and other explanations can be offered. An important one concerns shifts in advertising. Advertising tactics, as noted earlier, moved toward "softer" themes, with greater emphasis on emotional appeals and on pleasurable rather than practical uses of the product. They also focused increasingly on women as primary consumers, and women were later associated with telephone sociability.[59] AT&T executives may have been late to adopt these new tactics, in part because their advertising agency, N. W. Ayer, was particularly conservative. But in this analysis, telephone advertising eventually followed general advertising, perhaps in part because AT&T executives attributed the success of the automobile and other technologies to this form of marketing.[60]

Still, there is circumstantial and direct evidence to suggest that the key change was the loosening, under the influence of public practices with the telephone, of the telegraph tradition's hold on the telephone industry.

■ Conclusion

Today, most residential calls are made to friends and family, often for sociable conversations. That may well have been true two or three generations ago, too.[61] Today, the telephone industry encourages such calls; seventy-five years ago it did not. Telephone salesmen then claimed the residential telephone was good for emergencies; that function is now taken for granted. Telephone salesmen then claimed the telephone was good for marketing; that function persists ("Let your fingers do the walking. . . .") but never seemed to be too important to residential subscribers.[62] The sociability function seems so obviously important today, and yet was ignored or resisted by the industry for almost the first half of its history.

The story of how and why the telephone industry discovered sociability provides a few lessons for understanding the nature of technological diffusion. It suggests that promoters of a technology do not necessarily know or determine its final uses; that they seek problems or "needs" for which their technology is the answer (cf. the home computer business); but that consumers may ultimately determine those uses for the promoters. And the story suggests that, in promoting a technology, vendors are constrained not only by its technical and economic attributes but also by an interpretation of its uses shaped by its and their own histories, a cultural constraint that can be enduring and powerful.

---

[59]Recall that, early on, women were associated in telephone advertising with emergencies, security, and shopping.

[60]On changes in advertising, see sources cited in n. 21 above. The comment on N. W. Ayer's conservatism comes from Roland Marchand (personal communication).

[61]It is difficult to establish for what purpose people actually use the telephone. A few studies suggest that most calls by far are made for social reasons, to friends and family. (This does not mean, however, that people subscribe to telephone service for such purposes.) See Field Research Corporation, *Residence Customer Usage and Demographic Characteristics Study: Summary*, conducted for Pacific Bell, 1985 (courtesy R. Somer, Pacific Bell); B. D. Singer, *Social Functions of the Telephone* (Palo Alto, Calif.: R&E Associates, 1981), esp. p. 20; M. Mayer, "The Telephone and the Uses of Time," in Pool, *Social Impact* (n. 2 above), pp. 225–45; and A. H. Wurtzel and C. Turner, "Latent Functions of the Telephone," ibid., pp. 246–61.

[62]A 1934 survey found that up to 50 percent of women respondents with telephones were "favorable" to shopping by telephone. Presumably, fewer actually did so; see J. M. Shaw, "Buying by Telephone at Department Stores," *Bell Telephone Quarterly* 13 (July 1934): 267–88. This is true despite major emphases on telephone shopping in industry advertising. See also Fischer, "Women" (n. 41 above).

Shaping the Early Development of Television

Reading 11

Jan van den Ende, Wim Ravesteijn, and Dirk De Wit

Few of the billions of today's television viewers are aware that their early predecessors were served by a mechanical form of television. The central element in this early television was a rotating disc with a spiral series of perforations, called the Nipkow disc, after its Russian inventor Paul Nipkow. Two such discs were needed, one in the camera and one in the television itself. Rotating Nipkow discs, lamps, and light-sensitive elements together formed a means of recording and displaying television images.

Today the mechanical television is usually regarded as a historical curiosity; a cumbersome machine with a large, motor-driven wheel, a window, and flickering, blurred images: at best, merely a technological precursor to electronic television. However, we believe that there were once important incentives supporting mechanical television in its rivalry to electronic television. The electronic television gained preeminence not because of purely technical considerations, as is so often assumed, but also because of social ones. The programming that the television public wanted to see (the "software") influenced the choice in television technology (the "hardware"), and the institutional incorporation of television also played an important role.

Received wisdom on the early history of television originates from writers who have largely focused on technical aspects. This material conforms to the traditional approaches to the history of technology, which explained the development of technology mainly referring to technical factors and the ideas of inventors. We will address these approaches with the usual term "internalist," although it has been demonstrated that this concept has never been defined accurately, in contrast to its opposite "externalist" [2]. The internalist kind of historical study resulted in good narratives but offered little explanation for the direction of technological developments. The internalists assumed that the new technology was always superior to the old and took this as the self-evident reason for the spread of new technologies within society. In the case of television, according to this view mechanical television may well have enjoyed some success, but did not stand a chance against the newer, and therefore better, electronic television. Naturally, in a short while electronic television had gained complete superiority and mechanical television could be regarded as a historical curiosity. In hindsight, those involved could also be appraised accordingly: conservatives had held on to the mechanical system, those who had embraced electronic television early on were progressives.

It has since become clear that the internalist approach provides an insufficient explanation of technological development. For example, it gives no explanation of the fact that new technologies sometimes lie idle for years before coming into use, while others quickly displace the old. With regard to the history of television, an internalist approach fails to answer two questions. The first is, why was mechanical television so successful? Around 1930 the system was in widespread use in both Great Britain and the United States. In America, numerous broadcasting stations were in operation; in England, the BBC broadcasted regular programs. For many years the system underwent continuous development and in 1936, when an extensive comparison between the mechanical and the electronic systems was organized in England, it was still not clear which was better. The second question is, why did the development of television technology run such different courses in the United States and in Great Britain? In the U.S., development was discontinuous; around 1930 a "television boom" saw the use of mechanical television grow strongly, then stagnate around 1933. Electronic television subsequently came slowly into more widespread use. In the U.K., the development was more gradual; the use of mechanical systems continued right up to a national comparison in 1936, after which they were replaced by electronic systems.

Besides the internalist approach, a second type of historical perspective on the growth of television is provided by media history. In this, the social infrastructure necessary for television, such as broadcasting organizations, actors, and press offices, is held to be central. Media historians also pay attention to programming and to public evaluation. However, the television itself is kept out of the discussion; in practice it is regarded as a "given" technology, the implementation of which (in the form of television programs) is socially determined.[1] Media historians therefore regard the choice between mechanical and electronic television systems as a technical issue. Nevertheless, media historians would throw more light onto the subject were it not for the fact that they are primarily concerned with the later, rather than the earliest, stages in the development and spread of television.

Despite all criticism on internalist and other traditional approaches of technology development, they persist in recent literature, while new social-constructivist approaches, until now, have not led to a satisfactory alternative theoretical framework.[2] Basically, the new approaches stress the interaction between internal and external (or contextual) factors in technology development. We support this notion in regard to the early development of television. We do so on the basis of materials largely extracted from writers who focused on the technical history of television.[3] We see that it may be difficult to demonstrate the role of contextual influences in the history of television, but hope our analysis contributes in indicating that such influences indeed existed, even on the basis of existing materials.[4]

We focus on the choice between mechanical and electronic television systems. We analyze this choice from a broad perspective, with attention to social and technical factors and their interaction. We employ social-constructivist notions, especially the insight that several choices are always present in the historical development of a technology (see [6], [26]). Furthermore we take upon the idea that social groups, according to the meanings they attribute to the technology, can influence the course of this development.[5] We understand social groups, not on the social-constructivist basis of the meanings attached to a given technology, but in the more traditional sociological sense. We also include individuals, organizations, and the relationships between them in our deliberations (see also, [29]).

In contrast to social-constructivism, we accord a specific role to technological factors. In our approach, a given technology has a certain objective reality that circumscribes the meanings that can be attributed to it (see also [30]). This is one of the reasons that agreement often exists between different parties on the given attributes of a new technology, even when they have different preferences with regard to the technology itself. We shall see that both mechanical and electronic television had strong and weak points that were recognized by most of those involved. Moreover, we hold that to a certain extent technological innovations do indeed have their own dynamic. Physical reality puts constraints and offers opportunities for technological development. For instance, technical limitations in 1930s Britain made it very difficult to develop a mechanical camera for outside broadcasts. Moreover, innovations developed for a specific application create opportunities in other fields. The invention of the vacuum tube, which made the development of electronic television possible, forms an example.

We concentrate our attention on developments as they took place in two of the leading countries in this field, Great Britain and the United States, with only a passing comment on a third front runner, Germany. The Netherlands provide us with an illustration of developments in a "subsequent" country; although the choice between the two television systems was strongly influenced by events abroad, the contest between mechanical

---

[1]This standpoint is explicitly embodied in [3].

[2]Even one of the main protagonists of social constructivist approaches, Wiebe Bijker, states this in. [4].

[3]There are some exceptions of publications that focus on the social context, for instance J. Udelson's, *The Great Television Race, a History of the American Television Industry 1925–1941* [5].

[4]Citing M.J. Mulkay, Wiebe Bijker *et al.* explicitly state that it is difficult to apply contextual approaches on the history of television. See [6]. See also [7]–[10].

[5]Our approach shows similarities to Patrick C. Carbonara who also focuses on the circle of interdependence between players that affected the dispersion of monochrome and color television in [28].

and electronic television was also fought on Dutch soil. In fact, the unique social situation in the Netherlands gave developments a specific character.

■ Early History

The nineteenth and first half of the twentieth century saw a large number of new ideas and inventions having to do with sending images over long distances. In hindsight, many of these can be seen as having formed part of the history of television; however, the inventors often had other applications in mind, such as a "picture telegraph," film, or "picture telephone."

A number of historical moments come immediately to mind [6]. Firstly, the invention of the selenium cell, whose photo-electric characteristics were accidentally discovered by an English telegrapher in 1873. In 1875, the American G.R. Carey was the first to suggest that it might be possible to send images by electricity: he talked of "seeing by electricity", in analogy to "hearing by electricity" with a telephone. The telephone inventor, Alexander Graham Bell, also thought that an image could be added to the sound.

At that time, researchers saw the main problem as one of being unable to send an entire image over just one line. Between 1880 and 1925, therefore, various scanning devices were proposed to allow the image to be sent, piece by piece, over one line. One of the most important of these was devised by Paul Nipkow, whose invention formed the basis for many subsequent mechanical television systems. While still a student at the University of Berlin in 1884, Nipkow had come up with the idea of a television that employed rotating perforated discs. The camera would consist of such a disc, a number of lenses, and a selenium cell. It would work as follows: a light source illuminated the object; lenses projected the reflected light onto the rotating disc; the disc allowed one point of the projected image at a time onto the selenium cell, so that for each complete rotation of the disc the entire image had been projected, a bit at a time, onto the selenium cell; the cell transformed this varying light intensity into an electrical current, and thereby turned the image into a series of consecutive electrical impulses. In the television receiver, the process would be reversed: the electrical signal would be amplified and connected to a lamp situated behind a synchronously rotating Nipkow disc, so that the lamp flickered with an intensity corresponding to the electrical signal generated by the selenium cell. The image would then be visible on the face of the disc. Nipkow never built a prototype: the first fully operational model based on his principles was hot produced until 1925.

Operation of mechanical television

Around 1910, the Russian Boris Rosing developed a television system in which the function of the Nipkow disc was performed by rotating mirrors. His television receiver, however, employed a cathode ray tube, developed in 1897 by Karl Ferdinand Braun at the University of Straatsburg. In this tube a glowing cathode radiated electrons which were then guided towards a light-emitting screen by means of an electromagnetic field. In 1911 Rosing succeeded in producing an image consisting of four lines using this method. Rosing called his system the "electric telescope." Rapid developmental progress was prevented by problems in directing the cathode rays. Tragically, Rosing himself was arrested in the turmoil of the Russian revolution, exiled, and never heard of again.

At about the same time as Rosing, the Englishman A. A. Campbell Swinton devised a completely electronic television system. In recording, the image was projected onto a photo-electric surface within a vacuum tube, and an electron beam scanned this surface. Campbell Swinton never succeeded in building a working model, but his work was to form the basis of the first patent application for a fully electronic television in 1929.

The following years also saw improvements in many of the elements of the mechanical system. In 1913 the German physicists Julius Elster and Hans Geitel made progress with the selenium cell and in 1917 D. McFarlan Moore developed a neon lamp capable of the rapid flickering required of the light source in the television receiver.

So by 1920 various ideas existed for television systems. Designs existed for a system using mechanical cameras and receivers (using Nipkow discs), a system employing a mechanical camera (with rotating mirrors) and an electronic receiver, and a system employing both an electronic camera and an electronic receiver. None of these systems could yet boast of a working prototype; but this was to change fast.

Mechanical Television in Great Britain: Baird

Thanks to the work of John Logie Baird, Great Britain can be called the birthplace of mechanical television.[6] Baird was trained as an electrical engineer at the Royal Technical College in Scotland and was a born innovator. After a number of failures, which included a jam factory in Trinidad, in 1923 he began developing a mechanical television system based on Nipkow's principles [10]. Within a year he had succeeded in sending shadowy images from a camera to a television receiver along an electrical wire. In April 1925, Baird gave the first public demonstration of his television system. The television produced eight image lines. It was a primitive affair; the Nipkow discs were made of hatbox lids.

Nevertheless, the demonstration was a success. A London *Times* journalist described the images as being faint and blurred but nevertheless sharp enough to be able to distinguish facial expressions. Baird was able to obtain financial support from a number of sources which enabled him to continue his research. He formed Baird Television Ltd. and in 1926 was able to present his new "televisor" to members of the Royal Institute arid *The Times*. The pictures were still of a person's head and shoulders, but Baird was now working with thirty lines; a neon lamp illuminated the disc in the television, and the discs in the camera and the receiver were electrically synchronized [11]. Moreover, he now used a new lighting method, the "spotlight" method, in which he changed the positions of the light source and photocells. The light source was behind the disc and the photocells in front, trained on the object. It was the object that was illuminated piece by piece. By employing more photocells, the total effective light intensity could be raised and this method also prevented the object from being excessively strongly lit.

There were three main technical problems. Baird himself complained constantly about the quality of the lenses and the photocells, his employees pointed to the imperfect signal amplification, and according to spectators the mechanical aspects, the cardboard discs, the wires and strings were shoddy. In an attempt to improve the working of the photocell he even considered using a human eye; a surgeon obtained one for him but after a crude dissection with a razor he gave up and threw it into a canal.

Baird wanted to provide an integrated system, that is, he wanted to provide programs as well as equipment. He was not interested in collaborating directly with the national broadcasting company, the BBC. He wanted to finance his research out of the sale of television sets and his programs out of advertising. However, the organization responsible for controlling broadcasting, the General Post Office (GPO), objecting to the low image quality, refused Baird permission to use his transmitters. In the meantime, he devised a number of remarkable firsts, for example, the *Noctovisor* (1926), an invention employing infrared light which ensured, as had the "spotlight" method, that actors were not subjected to excessive lighting intensities. With transatlantic transmission (1923) and color television (1929) Baird succeeded in attracting publicity. He needed it badly; his good contacts with supportive patrons were vital, since as long as he could not transmit and sold no television sets, Baird had no other income.

In 1929, Baird finally received permission to employ a medium-wave radio frequency for his transmissions. He worked with thirty lines and 12.5 images/s. The reason for the GPO's change of mind was that television had now become an issue of national prestige, and Baird had begun to establish international contacts (he also worked, for instance, with the German Fernseh AG) and had threatened to move to the United States if he did not receive a British transmission license. Though he marketed his television sets, or *televisors,* himself, they were manufactured by a number of other companies and were not immediately available; it would be six months before they would be on the market. In this first period the transmissions were mostly intended for amateur enthusiasts. Two radios were needed, one tuned to the audio signal, the other to the image signal; the loudspeaker output of the second formed the input signal for a homemade television receiver.

Baird's television sets cost £25, an average monthly wage. The DIY kit versions that he also marketed cost between £12 and £16 pounds. All models had a screen measuring 5 cm by 10 cm. Estimates of the number of television sets in use varied widely. In January 1931 Baird reported to the GPO that he had sold 1000, but at the end of 1932 he stated that 500 were in use. An unknown number of homemade sets were also sold. In addition, other companies, such as Electrical Music Industries (EMI), began to develop mechanical television systems.

---

[6]For the history of television in Great Britain, we have made extensive use of [9].

In 1933 the BBC put the figure of television sets in use at 8000. Slowly but surely, mechanical television was becoming a success.

Baird was now working on improving the image quality by increasing the number of image lines, but this was no simple matter. It meant increasing the diameter of the Nipkow disc, increasing the number of its perforations and reducing their size. In 1933, Baird managed to increase the number of lines to 120, and in 1936 to 240. By this time the perforations were so small that dust was blocking them and so Baird switched to a camera employing revolving mirrors. He also brought larger, more expensive television sets onto the market, with a screen measuring 10 cm by 22 cm; these varied from £50 to £75 in price. Few if any of these sets were ever sold.

Electronic Television in Great Britain: A Gradual Takeover

The development of electronic television in the U.K. made use of American inventions and patents. One organization, in particular, was involved: the specially-formed company Marconi-EMI, of which the aforementioned EMI was one of the founders. The new company was given access to diverse American patents, including one for an electronic camera, the *iconoscope*, developed for the American company RCA by the Russian Vladimir K. Zworykin. He had studied under Rosing in Russia. The heart of the iconoscope was a vacuum tube, in which the image was projected onto a light-sensitive surface. Each point on this surface formed a small capacitor together with the metal plate situated immediately behind it. An electron beam scanned the surface vertically and horizontally, and the discharge current, which varied with the degree of illumination at each scanned point, provided the television signal.

However, the range of the transmitters was smaller for electronic television. The reason for this was that an increased amount of information per second could be transmitted only on higher radio frequencies, particularly in the VHF band, whose range was only about 50 km. Mechanical television mostly employed short- and medium-wave frequencies which had much greater ranges. Of course, it was not necessary for electronic television to have larger number of image lines and a higher amount of information per second than mechanical systems. Electronic television could also be employed with similar numbers of image lines, and then the same frequencies could be used and the same geographic range be reached. However, in that case the potentials and advantages of the electronic system compared with the mechanical one would not exist, whereas the electronic system still had the disadvantage of a higher price. So in practice, the amount of information of electronic was indeed higher. The eventual solution for electronic TV broadcasts was the construction of relay stations that allowed the VHF signal to be heard further afield.

In England, Marconi-EMI developed a camera, the *emitron*, based on Zworykin's principles, and which the company displayed at the 1935 annual radio exhibition. The *emitron* had an improved image quality and was also mobile. A number of other television systems were also on show. The electronic television sets were more expensive than the mechanical ones; at about £100 they represented an average Englishman's three months' wages, but the price did nothing to dampen the public's enthusiasm. Only for films (where the object was relatively small and could be perfectly lit) was Baird's mechanical system still considered more attractive. Baird tried to improve his system further, and even started to do research into electronic television.

In 1936, the Television Advisory Committee, which comprised representatives of the GPO and the BBC, created a two-year period to decide which system was better. Numerous television system manufacturers put themselves forward but only Baird and Marconi-EMI made it to the final round. Baird only participated with mechanical systems in the contest. He was now working with 240 lines and had two recording systems. For close-ups and studio work he used a conventional mechanical camera. For outside broadcasts he had developed a special process: the object was first filmed, the film automatically ran on through a developing machine, and the developed film was then broadcast using a special scanner. The whole process took only forty seconds. The purpose was to circumvent the mechanical camera's inadequacy in weak light. Baird's mechanical outside-broadcast camera had important disadvantages; it was large and difficult to transport, and the fragility of the film only increased the chances of an accident. Moreover, the whole process was extremely expensive. Baird's troubles took a turn for the worse when Crystal Palace burnt down in 1936; his company was housed there and much of his equipment was lost.

EMI was already demonstrating the emitron when the two-year trial period was introduced. The machine used 405 lines, and the television screen was now 30 cm by 30 cm. The screen tube was so long that it was mounted vertically, with a 45° mirror providing the final viewing screen. Even before the trial period was over, the machine was declared the best available; the image quality was better, the system was relatively reliable, and the emitron camera, being truly portable, was better suited to outside broadcasts than were Baird's mechanical cameras.

In January 1937 the BBC—which the GPO had accorded a television broadcasting monopoly, as it had in radio broadcasting—began transmitting regular broadcasts using an electronic television system with 405 lines. At first, few television sets were sold but a month later EMT dropped the price from £100 to £60. Six months after broadcasts had begun, 1500 sets (of all makes combined) had been sold. The spread of electronic television had begun.

■ Social Shaping

The development of mechanical television in the U.K. involved a number of different social groups and organizations: the manufacturers (Baird and his company), the supervising governmental body (the GPO), the broadcasting company (the BBC), actors, amateur television enthusiasts, and the viewing public. Each group attributed different meanings to television.

For Baird, mechanical television was a means to set up and run a company. Television, for him, was a marketable product; his desire to produce his own programs was part of his marketing strategy. This was made possible in 1929 when the BBC first allowed him to use a transmitter. His broadcasts were mostly made up of studio entertainment, educational programs, and films. In this period, while he was broadcasting on a small scale and sold no television receivers, he depended on financiers to sponsor his activities. This makes clear why he put so much effort into coming up with new inventions; they would convince his patrons of the viability of the medium.

To start with, the BBC showed extreme caution; the broadcasting company would only be interested in talks about transmission, let alone participation, when an acceptable image quality could be guaranteed. Cinema film quality was informally taken as a standard. The difficulty of making outside broadcasts with the mechanical system also formed a problem for the BBC. Baird's solution to this problem, using an intermediary film stage, had its own problems; the camera was practically immobile, and the time delay that the film development produced was another possible problem.

At first the GPO was reluctant to give Baird a broadcasting license. Their argument was the same as the BBC's: Baird's inability to guarantee a satisfactory image quality. Only when Baird threatened to leave the U.K. and set up in business abroad was he finally given a license—for England could not afford to lose Baird's company; television had become an issue of national importance.

Actors, too, had problems with mechanical television, for it depended on intolerably intense studio lighting. Baird's "spotlight" method was only a partial solution to this problem; the process put the actors into flickering light with the rest of the studio in utter darkness, so that they could read their parts only with difficulty and could not see the director. The system also left the actors precious little freedom of movement. At one point the BBC feared that actors would refuse to extend their cooperation to any mechanical television broadcasts.

Electronic television promised to solve most of these problems, but only by introducing a new problem, that of the extremely limited range of the existing VHF transmitters; many viewers would be left without a signal. To tackle this problem, an entire network of new VHF transmitters had to be set up. In 1934, when the BBC was considering replacing the mechanical system with the electronic, letters arrived from listeners in Scotland and the north of England, protesting that the change would leave them with no television signal at all. Electronic television, they argued, would divide British television viewers into two geographical halves: those within reach of VHF transmissions, and those without. The price of electronic television sets was probably another obstacle to the consuming public; they were much more expensive than the mechanical model.

Amateur television enthusiasts also found difficulties with the electronic version. VHF frequencies could not be picked up on ordinary radios, and the electronic television was too complex to build at home. And finally, electronic television of course brought Baird a big problem: it formed a threat to the continuation of his company.

Mechanical Television in the United States: A "Boom"

Baird had an American counterpart. It was another inventor, Charles Francis Jenkins, who gave America its first mechanical television.[7] In June 1925 he was able to show pictures of a Dutch windmill turning; the images had been taken from a film screen. Jenkins' camera used two ground glass discs, that formed prisms, which deflected the beam of light in one direction as the discs rotated. He called his system *radiovision* because the system was wireless, preferring to reserve the term *television* for transmissions that took place over electrical cables, in analogy with the telephone and the telegraph. He envisaged numerous applications, such as film shows, theater shows, and sporting contests [5]. Like Baird he formed a company, but unlike Baird's it was supported by the banks.

It was not long before other large concerns began to show interest. AT&T's Bell Laboratories started work in this field in 1925 and General Electric also carried out research into mechanical television around this time. AT&T saw the device's potential as a "picture telephone," and between 1925 and 1929 the company poured over $300 000 into research into mechanical television. Two hundred engineers and scientists were put to work. By 1927 Bell was able to demonstrate two systems. One employed a large screen measuring 60 cm by 75 cm, the other a small screen measuring 5 cm by 6 cm. The latter was intended for installation in a "picture phone." In June 1929, exactly four years after Jenkins' original television, AT&T demonstrated a mechanical color television, which worked using three color filters and three photocells, each photocell being connected via one of three transmission cables to a separate neon lamp in the television receiver.

From 1928 to 1932, mechanical television in the USA prospered — these were the years of a "television boom." Dozens of television stations started broadcasting, having been granted "experimental broadcasting licenses" by the Federal Radio Commission (FRC), the body that had been set up to regulate the use of radio waves. In 1931, 31 television stations were in active service and Chicago alone had an estimated 8000 television sets. Thereafter most of these stations disappeared, except for three from universities in the Chicago area that continued to broadcast educational programs using the mechanical system until 1937. Although the depression may also have been of influence, after this initial period it gradually became clear that the mechanical television system was not a viable option in the United States. We shall shortly examine the reasons below. Jenkins' company went bankrupt, and other manufacturers had it no easier.

Electronic Television in the U.S.: A Slow Start

The first patent application for an entirely electronic television system came from the above-mentioned Zworykin. In the 1920s he worked at Westinghouse in the U.S., where at first little interest was shown in his activities but where he was later given a free hand. In 1923, Zworykin applied for a patent on his device, although the machine had yet to produce more than the shadowy image of a cross.

Six years later Zworykin was able to present a working cathode ray tube television set. Around this time he went to work at RCA, a company that had set up a central radio research laboratory in which researchers from various RCA companies were at work. In 1930 Zworykin was made head of the laboratory, and under his direction in 1933 it produced the iconoscope, the electronic camera.

The higher image quality of electronic television was due, in Zworykin's terms, to the "storing of electrical charges." In the mechanical camera, an increase in the number of screen image lines reduced the amount of light falling onto the selenium cell at a given moment; since the cell's low-light sensitivity was limited, at a certain point this began to create problems. In the electronic camera, the light sensitive plate was permanently illuminated, and in raising the number of screen lines, only the scanning speed was increased [13]. This allowed the electronic camera to operate in lower light levels than the mechanical camera.

Zworykin's successes encouraged RCA to invest a million dollars in television in 1935 alone. The concern built a new television transmitter on the Empire State Building in New York City and a special TV studio was built at the National Broadcasting Company (NBC), the RCA broadcasting organization. Although the company still had no official license to transmit TV commercially, in 1936 RCA began a campaign to attract potential TV advertisers. RCA's efforts to gain advantage over its competitors by building a patent portfolio became obstructed by the efforts of one man, who started with a single invention, Philo T. Farnsworth.

---

[7]For the history of the mechanical television in the United States we have made extensive use of [7] and [12].

Farnsworth was responsible for the invention of another electronic television system. Farnsworth started working on what he called his television system in the mid 1920s. He managed to get several Californian businessmen, especially George Everson, interested in his ideas. In the fall of 1927 Farnsworth succeeded in a laboratory demonstration. Within a year his television system was made public with an article in the *San Francisco Chronicle* describing Farnsworth' camera, the *image dissector*, as ordinary jars that housewives used to preserve fruit. Farnsworth' achievements did not go unnoticed. In 1930 RCA tried to acquire "Television Laboratories Inc.," an offer that was declined by George Everson. After a brief liaison with Philco, Farnsworth continued on his own account with the Farnsworth Television Inc. The company soon became entangled in a patent conflict with RCA, which argued that Zworykin's iconoscope was the first camera.

Commercialization of television in the U.S. was delayed by the fight over patents. Farnsworth could not license his patents in the U.S. as long as his patents were under contest. In 1934 Farnsworth received an invitation from Baird Television for a patent license in England. Baird was at that stage forced by his financial backer, British Gaumont, to turn his mechanical television into an electronic system, because BBC had advised him to stop working on his experiments. With the money from this license Farnsworth could continue the perfection of his television system.

Electronic TV broadcasts took place in the U.S. on a limited scale after 1932. A number of stations that had employed mechanical equipment continued with electronic. RCA, which had broadcast with mechanical and shortly with a hybrid mechanical/electronic system, in 1933 set up a network of transmitters for all-electronic television. In this way the company offset the limited VHF signal range. In 1935 AT&T produced the coaxial cable, the first of which was laid between New York and Philadelphia. Nevertheless, electronic television made slow progress over the next few years. This was partly caused by the fact that companies such as Philco and Dumont, which were not yet able to produce a complete system themselves, convinced the Federal Communications Commission (FCC), the successor of the FRC, that no consensus yet existed within the industry on an "ideal" television system. It was not until Farnsworth entered into a cross-licensing agreement with AT&T in 1937 that could have left RCA out of commercial television altogether, that RCA and Farnsworth began the negotiation of cross-licensing agreements. In 1939 RCA was obliged to reach a licensing agreement with Farnsworth before they could market their television. In the meantime the agreement between AT&T and Farnsworth convinced the Columbia Broadcasting System (CBS) that commercial television had become viable.

In 1939 television became "official" when President F.D. Roosevelt initiated the first television broadcast at the New York World's Fair. In the next week the first television sets went on sale. Commercial television was within reach, but the World's Fair only started a new phase of experimentation in which the public was to be included through the sale of a handful of receivers. Not even RCA had permission to sell commercial time to advertisers until the FCC decided on standardization. However, the FCC delayed standardization since it demanded unanimity within the industry. FCC's concern for consensus became the catalyst for the Federal Government to establish the National Television System Committee (NTSC), which operated under the auspices of the branch organization of the industry, the Radio Manufacturer's Association. The NTSC reached consensus soon after the outbreak of World War II, and in 1941 standards were established. But now the war would prevent the commercialization of television.

■ Social Shaping of Television in the United States

The U.S. media infrastructure differed greatly from the British, a situation dating from the radio age. In the early years of radio development, various companies energetically competed for and contested patent and broadcasting rights. In setting up NBC in 1926, RCA brought a number of these companies together; the radio stations belonging to the RCA, General Electric, Westinghouse, and AT&T were now operated by a single organization. AT&T was now responsible only for the transmission itself.

As had also been the case for radio, mechanical television equipment was produced by a number of different manufacturers, including Jenkins, AT&T, and General Electric. A number of these companies also undertook broadcasting activities; one of these was the RCA's own NBC. Most of the manufacturers saw mechanical television as a broadcasting medium, but AT&T concentrated on the potential for a "picture phone."

During the radio age, government and industry had reached an agreement embodied in the Radio Act of 1927. Government's primary aim had been to prevent the formation of monopolies in the broadcasting, programming, and television manufacturing industries. Secondly, it sought standardization, so that every radio could receive every station. It also expressed the view that the development of radio should remain a matter of private initiative; radio manufacturers and broadcasting companies should finance themselves. The governmental Federal Radio Commission was formed to confer licenses and to allocate radio frequencies, but was not to concern itself with the canvassing of advertising, nor with most aspects of programming. With the subsequent introduction of television, the FRC — later to become the FCC — was also the regulatory body.

Television broadcasts were dominated by singing, cabaret, educational programs (which included, for instance, a course on bookbinding) provided by the universities, and the occasional film or sporting contest. In Chicago a television station was run by a newspaper together with a television manufacturer. It televised little news, though; most of the programming consisted of cartoons.

In the U.S., too, amateur television receivers formed a significant social group. Special DIY televisions, which were considerably cheaper than the electronic sets, were brought onto the market.

What specific problems were faced by the various social groups using mechanical television in the U.S.? The FRC was faced with three problems; first, it was unsatisfied with the image quality and therefore only conferred temporary licenses. Secondly, there was still no transmission standard. Different stations employed different standards with regard to the number of lines and the number of images per second. In choosing between different standards, television buyers ended up effectively choosing between stations. Later the FCC therefore advocated standardization. Thirdly, station frequencies regularly overlapped. Here, too, the FRC wanted to sort things out.

For the broadcasting companies, the main problem was one of income. A temporary license did not allow a station to broadcast advertisements, and this denied them an important source of income. A third social group, the actors, had the same difficulties with the lighting as they did in the U.K. Finally, for the consumer public, the absence of technical standards and the overlapping of broadcast frequencies were the greatest problems.

How did this compare with the situation for electronic television? For the most important manufacturers of electronic television (RCA, Farnsworth, and Philco) patent rights formed a problem; as we have seen, RCA was eventually obliged to come to a licensing agreement with Farnsworth. The FCC, on the other hand, had to decide on certain technical standards that would unavoidably hinder the introduction of later technical improvements. Nevertheless, after the establishment of the NTSC, in 1941 a set of standards was agreed to. In other countries, such standards were only drawn up much later, and this explains why to this day the U.S. has a technically less sophisticated television system.

The broadcasters employing electronic television were by and large the same companies that had employed mechanical television. The main problem was the limited VHF transmission range. To begin with, reception was limited to the larger cities, and only after 1933 was a solution provided by a wider network of transmitters. By the end of the 1930s broadcasters and manufacturers joined forces after the FCC, in order to promote technical improvements, decided to confer broadcasting licenses only to those companies actively engaged in research into such improvements.

Since from the mid-1930s onwards broadcasters were allowed to carry advertisements, advertisers gradually became another important social group involved in electronic television. Initially, however, low viewing figures made these advertisers cautious.

The viewing public, meanwhile, was pleased that electronic television made it possible to broadcast events that took place outside the studio: sport and news. However, the absence of transmission standards was a problem for consumers, as in all probability was the expense of an electronic television set: in 1939 a set cost between $200 and $600. Again, part of the mechanical television viewing public was not reached by electronic television broadcasts. The conviction that areas outside the large cities should also have the benefit of television played a large part in the FCC's decision to allow the three universities that employed mechanical television to continue doing so for many years.

Finally, American amateur enthusiasts had the same problems with electronic television as did their British counterparts: its overwhelming technical complexity and the greater technical differences with ordinary radio reception.

■ Mechanical Television in the Netherlands: Hat Shows

For a short time, mechanical television enjoyed a certain popularity in the Netherlands, although this was less the case than in either the U.K. or the U.S.[8] Just as had been the case in these two countries, it was an individual pioneer, the Eindhoven radio amateur Freek Kerkhof, who got Dutch television off the ground. In 1924 he decided to develop a television system using Nipkow discs, but it was 1927 before he could demonstrate his ten-line system, and little more than shadowy images could be discerned. Philips followed developments in image transmission from 1925 onward. Between 1928 and 1931, radio and newspapers showed considerable interest in mechanical television.

Philip's first experiments date from this time. Although its research laboratory's directors had resisted mechanical television from the start, under commercial pressure they presented, in 1928, a television image made up of 48 lines, employing Nipkow discs, a photocell, and a neon lamp. The device was able to radiotransmit an image of static objects a distance of 400 m. Philips had two aims in mind for these demonstrations of mechanical television: first to show that it was interested in the medium, and second to demonstrate its shortcomings. For all its sophistication, the Nipkow television was a large, cumbersome and susceptible machine and Philips saw little future for it in Dutch households.

In the first half of the 1930s, Dutch broadcasters showed little interest in the medium of television. One, the Vereniging van Arbeiders en Radio Amateurs (VARA), was an exception; in 1931 it transmitted an experimental broadcast using a thirty-line machine obtained through the German television manufacturer Telehor AG. The 3 cm by 4 cm viewing screen was enlarged by means of an enormous lens. The broadcast lasted 15 min and was accompanied by widespread publicity. Radio magazines and newspapers described the limitations that thousands had witnessed for themselves. The portrait of a woman, they reported, had been indistinct and shaky; "one moment she was visible, and the next moment it looked more like a film of railway tracks shaken by the passage of a highspeed train" [18]. Not surprisingly, the public was warned not to expect too much of the new medium.

Kerkhof, in the meantime, continued his experiments. By 1935 he was able to demonstrate a thirty-line system, comprising a mechanical mirrordisc camera and three receivers, to the Dutch National Society of Radio Amateurs. The program showed Mrs. Kerkhof, busily displaying one hat after another. This demonstration was followed up in 1936 when Kerkhof started regular Sunday morning transmissions for Eindhoven amateurs using homemade 30-line receivers. A year later Kerkhof had built a second transmitter for the accompanying sound signal. He had also begun using cathode ray tubes in his cameras, which resulted in a considerably improved image contrast. After other improvements were made in the studios, he was able to start transmissions employing well-known artists, and these broadcasts continued until his transmission license was withdrawn as the result of Dutch war mobilization in 1939.

■ Electronic Television in the Netherlands: Philips and the Government

Philips began building an experimental electronic television system in the 1930s.[9] Their research laboratory experimented with camera recordings in the studio, outside broadcasts, and with film shows. In 1937 Philips began to give public demonstrations; systems were available using 405 and 567 lines, fn the Utrecht trade fair held in the spring of 1938, Philips demonstrated a number of 405-line television sets and also a large-screen set called the *protelgram*. The yellow-reddish image of the mechanical TV set had now given way to the sepia-greenish image of the electronic ones.

For all this, Philips was not actually very interested in television, a view the company retained right up to the 1940s. Philips directors saw the limited range of VHF transmitters and the high cost of television sets as serious obstacles to their successful marketing. Neither did they expect the public to be satisfied with a blurred image measuring only three by seven centimeters. And Philips did not expect the glass industry to suddenly come up with a large enough image tube.

Philips concentrated on two alternatives to television: the *huiscineac* and the above-mentioned *protelgram*, which was first shown at the Radiolympia in England in 1937. With the *huiscineac* (literally, "home cinema")

---

[8]A small number of Dutch publications have appeared on the development of mechanical and electronic television. Sources for the study of mechanical television are [14]–[16]; see also [17].

viewers could project films at home, but the process was ultimately never exploited commercially. With the *protelgram*, Philips was reacting to the fact that many people found the picture of the first generation receiver disappointingly small: the *protelgram* projected a small, intense electronic screen image, by means of a large spherical mirror and a correcting lens, onto a screen or wall [22]. The image was then 40 cm by 50 cm, but its clarity left much to be desired. In 1937 Philips sold several sets, but shortly afterwards was proposing to take them back because the tube life was unacceptably low.

Philips also suffered an unpleasant and unexpected side effect from its experiments with television; the "picture radio" led to a stagnation in demand for ordinary radios [17]! In 1938, Philips therefore embarked on a rather unusual campaign; staff traveled throughout Europe ostensibly demonstrating televisions but actually concentrating on promoting the sale of radios. For Philips too, the start of the World War II, put a lengthy stop to further activities [23].

The radio broadcasting companies had not ignored these developments. In 1935, three of these companies submitted an application to jointly run a television station. This application led to the formation of the Television Commission, which was to seek information and to advise on the technical, legal, and economic aspects of mechanical and electronic television systems. The existing administrative structure with regard to radio served as a guideline. The Commission was composed of broadcasters and the PTT, the Dutch Post Office, which had a role in the provision of communication cables.

The Commission examined the situation in various countries, and it watched experimental transmissions of the German Olympic Games. Its interim report expressed reservations about costs and program content, and the Commission felt that the difficulties were serious enough to warrant deferring a regular television service for the time being. However, the rapid course of developments abroad made an experiment with electronic television imperative, a conclusion which was repeated in the Commission's definitive report published in December 1937. Broadcasters and the PTT would take care of the technical side of such an experimental transmission; television receivers would be situated in public places, with the expectation that they would later appear in private houses. Philips declared itself prepared to supply apparatus, and after the broadcasters had acknowledged themselves prepared to make the programs and foot the bill for the transmissions, the Commission announced that everything was ready for a trial television service. At this point, World War II intervened and the experiment was postponed.

After the war, the various groups involved took matters up once more. Philips still had the American marketing of its own *protelgram* high on its agenda; the company was still convinced that television would spread only slowly in the Netherlands. Nevertheless by 1947 Philip's experimental research laboratory Natlab had produced a television transmitter, and its receiver was almost ready for mass manufacture. When a Philips manager in America sent a telegram strongly urging his colleagues to increase their television activities (the *protelgram* was all very well, but it looked like the future lay in television), Philips changed course [24].

The company was given a license to transmit within a radius of 40 km, and began experimental broadcasts to a select public. Three times per week Philips employees, local worthies, and radio dealers were treated to ninety minutes of television. Philips television receivers operated at a standard 625 lines, in contrast to France's 819 lines and Britain's 405 [25], [26]. Together with the PTT, Philips began to lobby government for permission to start more frequent broadcasts. One of its arguments was that a good domestic market was crucial to the sale of television sets abroad. However, party political and broadcasting companies' reservations held back the introduction of an experimental television service until 1951.

Social Shaping of Television in the Netherlands

In the Netherlands too, various social groups and organizations influenced the development of both mechanical and electronic television: television amateurs, broadcasting companies (especially the VARA), Philips, government, and the PTT. As elsewhere, mechanical television was most important to the amateur enthusiasts. Kerkhof called for their comments on the programming. Ultimately, however, they formed no important pressure group in defense of mechanical television, and this had much to do with Kerkhof's attitudes. Unlike Baird, Kerkhof did not try to wring commercial profit out of television but aimed only to provide the best possible quality given the means available to amateurs.

Broadcasting companies were quick to lay claim to television, perhaps in order to prevent the medium from being stolen by other organizations. The VARA's 1931 experimental broadcasts gave an important boost to the development of mechanical television, but as a whole the broadcasting companies' involvement in these experiments was short-lived. The same can be said of Philips; though the company began experiments with mechanical television in the 1920s, it appears that these experiments were not considered of great importance. Government involvement was represented by the formation of the first Television Commission, representing the broadcasting companies and the PTT, and which appears to have been equally unenthusiastic about mechanical television.

By and large, the same groups were involved in the growth of electronic television. Philips was still barely interested; when the company began demonstrating this equipment in the late 1930s the main purpose was to stimulate the sale of ordinary radios. Moreover, Philips was principally interested in a medium for the domestic reproduction of large-screen films, and was therefore more preoccupied with its *protelgram*. Still, its apparent disinterest in television ought not to be exaggerated; a significant portion of Philips' patent applications in 1934 and 1935 had to do with television.

In an important sense radio was the medium of the 1930s, and since the same parties were involved this was reflected in the development of both mechanical and electronic television. Practically everyone was content to wait and see what television had to offer; this also had much to do with the limited television programming of the time. The widespread opinion was that if television had anything to offer beyond adding images to a radio signal, it was the showing of films. Neither did Philips' attitude do anything to stimulate the further development of mechanical television.

After the Second World War the Netherlands opened an intense debate on the introduction of electronic television. Most of the parties who had been involved before the war reappeared and it was remarkable how many of them had changed their minds completely in the meantime. Philips' vacillation had been replaced by an almost aggressive approach, after the company had been put onto the scent by developments abroad. Having also gained the support of the PTT, Philips argued in terms of employment opportunities, export opportunities, and national prestige, and that these would first require a secure position for Philips in the Dutch market. The broadcasting companies, too, had modified their views; they were prepared, primarily from a protective standpoint, to work with experimental television broadcasts.

Government adopted no very clear position but did raise a number of financial and cultural objections. Political and broadcasting representatives (who in the Dutch social system were already strongly aligned) shared a certain hesitancy with respect to television which yielded to the pressure being generated by Philips, the PTT, and Nozema, a regulatory government installed office. Administrative structures that had evolved in the age of radio went on to determine the way in which television was used; indeed, in programming terms, television was initially little more than radio with pictures.

■ Social Factors an Influence

In considering the developments recounted here, the choice between mechanical and electronic television can be perceived as having been the result of a combination of technical and social factors. Technical factors had to do with the advantages and disadvantages of each system. Mechanical television was cheap and TV sets could be constructed by amateurs themselves, all sorts of technical extensions (such as the use of color, image recordings, and large screens) were easy to put into practice, and as a consequence of low information density, the transmitters could have a long range. However, the image quality was poor and the system was best suited to studio work and film shows; it proved impossible to develop a practical camera for outside broadcasts. Electronic television on the other hand had a high image quality and was well suited to outside broadcasts, but the technique was expensive and not within easy reach of amateurs. Moreover, electronic television could only exploit its advantages in image quality if a higher number of image lines was applied. This required transmission on a VHF bandwidth with a low geographical range.

The choice between these two systems cannot, however, be attributed to these technical factors alone. Social factors also played a role. Different meanings were attributed to television by the various agents in its development: some wanted to see it used for public broadcasts, others for "picture phones." Its use in broadcasting was itself subject to differing appraisals: some saw it as a medium for film shows, others saw its potential as a news broadcaster, and still others noted its educational possibilities. In a social process involving a number of different social groups and organizations, the use of television for broadcasting purposes became predominant. With regard to programming, live events, and sports contests in particular, seem to be considered important. This appears, for instance, in the use of the 1936 Olympic Games in Berlin as a test between the two television systems in Germany. Electronic television was more suited to the broadcasting of such live events, as good image quality was vital and these events normally took place outdoors.

However, had another set of connotations predominated, for instance if television had been a medium of studio recordings and video telephony alone, then mechanical television would have stood a better chance. Certainly, in such applications its image quality would not have reached that of the electronic television, but the difference would have been smaller in these applications, and its advantages in other respects would probably have improved its overall acceptance. Although mechanical television had the advantage of a head start on electronic television, its early success largely resulted from the fact that studio recordings and video telephone were originally perceived as important purposes. Had there been a need for transmissions with worldwide reception (a need which is now met by satellites), mechanical television would also have had advantages. It is possible that other interpretations also played a part in the eventual choice for electronic television; actors, we may recall, had many difficulties with the strong lighting required by mechanical television.

In other words: internalist historical accounts take the importance of the image quality criterion for granted. In contrast to this, our approach to the early history of television reveals that in the course of a social process, image quality and the possibility of making outside broadcasts became dominant criteria.

The difference between historical developments in Great Britain and the United States must be attributed to the financial structures of the broadcasting organizations in each country. In the U.S., broadcasting companies had to meet their costs by selling advertising. Since they were given only experimental transmission licenses which stipulated that no advertising could be carried, they were deprived of this income and many eventually folded. In Britain, the BBC had a more secure future and the development of mechanical television became part of the national interest. Another explanation for the difference may lie in the fact that American mechanical television was faced with more problems, thanks to the large number of broadcasting companies, the lack of standardization, and limited radio frequencies. The reason that American universities continued to use mechanical television for so long is that it suited their purposes reasonably well; they broadcast mainly studio programs.

While the Netherlands was no television pioneer, it was no slow starter either. Early on, a mechanical system was developed that remained in use until 1939. Later, Dutch television fell behind the field. Philips, the most important Dutch electrical goods manufacturer, was only moderately interested in electronic television, and until 1947 it was more concerned with the alternatives to television. Developments abroad finally persuaded Philips to change tack.

A note should be made of the relationship between television and its predecessor, radio. We have seen that the organizations, set up by government and industry in all three countries, responsible for supervising radio broadcasts, were also given (or assumed) responsibility for television. Television did not develop in a political or industrial vacuum. Significantly, the organizational structure of radio was simply grafted onto television, though the technical system chosen was one less compatible with radio as far as frequency and type of technology was concerned.

Overall, our broad, technical-social perspective gives a better understanding of the early history of television than earlier one-sided approaches. Our analysis reveals, for example, that screen image quality and the possibility of making outside broadcasts became dominant criteria in the course of a social process. However, our analysis also shows that we cannot rub out technical factors, as some new approaches tend to do, thus in fact introducing a new type of bias. Perhaps, researchers busy in constructing new approaches should not polish off the older ones too soon, but try to find more embracing perspectives.

References

1. J. Van den Ende, W. Ravesteijn, D. de Wit, "Waarom geen Nipkowschijf in elk huiskamer? De sociale constructie van televisie," *Het Jaarboek Mediageschiedenis*, vol. 5, pp. 131–161, 1993. (Published by Stichting Film en Wetenschap, Zeeburgerkade 8, 1019 HA Amsterdam, the Netherlands).
2. S. Shapin, "Discipline and bounding: The history and sociology of science as seen through the externalism-internalism debate," *History of Science*, vol. 30, pp. 333–369, 1992.
3. R. Williams, *Television, Technology and Cultural Form*. London: Fontana, Collins, 1974.
4. W.E. Bijker, "Sociohistorical technology studies," in *Handbook of Science and Technology Studies*, S. Jasanoff, G.E. Markle, J.C. Peterson, and T. Pinch, Eds. Thousand Oaks/London/New Delhi, 1995. p. 255.
5. J. Udelson's, *The Great Television Race, A History of the American Television Industry 1925–1941*. AL: Univ. of Alabama Press, 1982.
6. W.E. Bijker, T.P. Hughes, and T. Pinch, Eds., *The Social Construction of Technological Systems*. Cambridge, MA: M.I.T. Press, 1987, pp. 41–42.
7. A. Abramson, *The History of Television, 1880–1941*. Jefferson, 1987.
8. *The History of Television from the Early Days to the Present*. London, IEE Conf. Pub. 271, 1986.
9. R.W. Burns, *British Television, The Formative Years*. London: Peregrinus, 1986.
10. T. McArthur and P. Waddell, *The Secret Life of John Logic Baird*. London: Hutchinson, 1986.
11. J. Swift, *Adventure in Vision, the First Twenty-five Years of Television*. London: Lehmann, 1950.
12. L.W. Lichty and M.C. Topping, Eds., *American Broadcasting*. New York: Hastings House, 1975.
13. G. Shiers, Ed., *Technical Development of Television*. New York: Amo, 1977.
14. W.J. de Gooijer, *Beheersing van technologische vernieuwing*. Alphen a/d Rijn, 1976.
15. J. Libbenga, "Tv or not tv, veertig jaar Nederlands televisie," in *Intermediair*, vol. 39, pp. 38–43, 1991.
16. G. Bekooy, *Philips honderd, een industriële onderneming*. Europese bibliotheek, 1991.
17. Jan Wieten, "Vuistslagen op peluws, de valse start van de televisie in Nederland," in *Het Jaarboek Mediageschiedenis* vol. 5, pp. 163–197, 1993.
18. E. Smulders, "Het Wonder van Morgen, de televisierage in Nederland 1928–1931," Rotterdam, Master's thesis, 1993, p. 56.
19. P. Gros, *Televisie, Parlement, Pers, Publiek*. Assen: Van Gorcum, 1960.
20. H. Wijfjes, *Hallo hier Hilversum, driekwart eeuw radio en televisie*. Weesp, 1985.
21. W. Knulst, *Van vaudeville tot video*. Alphen a/d Rijn, 1989.
22. K. Geddes and G. Bussey, *The Setmakers, A History of the Radio and Television Industry*. London: Brema, 1991.
23. L. van der Linden, "Vijftig jaar kastje kijken," in *De Ingenieur*, vol. 9, pp. 11–19, 1985.
24. H. Beunders, "De prehistorie van de televisie," in *NRC Handelsblad*, Sept. 28, 1991.
25. F. Kerkhof and W. Werner, *Televisie*. Deventer, 1951.
26. N.Tj. Swierstra, *Tien jaar televisie in Nederland*. Eindhoven: Philips, 1958.
27. W.E. Bijker, *Of Bicycles, Bakelites and Bulbs. Towards a Theory of Sociotechnical Change*. Cambridge, MA, 1995.
28. P.C. Carbonara, "A historical perspective of management, technology and innovation in the American television industry," Ph.D. diss., Univ. of Texas, Austin, 1989.
29. D. de Wit, *The Shaping of Automation, a Historical Analysis of the Interaction Between Technology and Organization, 1950–1985*. Hilversum, pp. 53–56, 1994.
30. J. van den Ende, *The Turn of the Tide. Computerization in Dutch Society, 1900–1965*. Delft, 1994.

The Personal Computer, 1972–1977

Reading 12

Paul Ceruzzi

Ready or not, computers are coming to the people.
That's good news, maybe the best since psychedelics.

Those words introduced a story in the fifth anniversary issue of *Rolling Stone* (December 7, 1972). "Spacewar: Fanatic Life and Symbolic Death Among the Computer Bums" was written by Stewart Brand, a lanky Californian who had already made a name for himself as the publisher of the *Whole Earth Catalog*. Brand's resumé was unique, even for an acknowledged hero of the counterculture. At Stanford in the 1960s, he had participated in Defense Department-sponsored experiments with hallucinogenic drugs. In 1968 he had helped Doug Engelbart demonstrate his work on interactive computing at a now-legendary session of the Fall Joint Computer Conference in San Francisco. Brand was no stranger to computers or to the novel ways one might employ them as interactive tools.

Brand was right. Computers did come to the people. The spread of computing to a mass market probably had a greater effect on society than the spread of mind-altering drugs. Personal computing, however, did not arrive in the way that Brand—or almost anyone else—thought it would. The development of personal computing followed a trajectory that is difficult to explain as rational. When trying to describe those years, from 1972 through 1977, one is reminded of Mark Twain's words: "Very few things happen at the right time, and the rest do not happen at all. The conscientious historian will correct these defects." This chapter will examine how computers came "to the people," not as Twain's historian would have written it, but as it really occurred.

What triggered Brand's insight was watching people at the Stanford Artificial Intelligence Laboratory playing a computer game, Spacewar. Spacewar revealed computing as far from the do-not-fold-spindle-or-mutilate punched-card environment as one could possibly find. The hardware they were using was not "personal," but the way it was being used was personal: for fun, interactively, with no concern for how many ticks of the processor one was using. That was what people wanted when, two years later, personal computers burst into the market.

Spacewar was running on a PDP-10. In terms of its hardware, a PDP-10 had nothing in common with the personal computers of the next decades. It was large—even DEC's own literature called it a mainframe. It had a 36-bit word length. A full system cost around a half million dollars and easily took up a room of its own. It used discrete transistors and magnetic cores, not integrated circuits, for logic and memory. Still, one can think of the PDP-10 as an ancestor of the personal computer. It was designed from the start to support interactive use. Although its time-sharing abilities were not as ambitious as those of MIT's Project MAC, it worked well. Of all the early time-sharing systems, the PDP-10 best created an illusion that each user was being given the full attention and resources of the computer. That illusion, in turn, created a mental model of what computing could be—a mental model that would later be realized in genuine personal computers.

Chapter 5 discussed the early development of time-sharing and the selection of a General Electric computer for Project MAC at MIT. While that was going on, the MIT Artificial Intelligence Laboratory obtained a DEC PDP-6, the PDP-10's immediate predecessor, for its research. According to the folklore, MIT students, especially members of the Tech Model Railroad Club, worked closely with DEC on the PDP-6, especially in developing an operating system for it, which would later have an influence on the PDP-10's system software. As a pun on the Compatible Time Sharing System that was running on an IBM mainframe nearby, the students called their PDP-6 system ITS—Incompatible Time Sharing System. The PDP-6 did not

"The Personal Computer, 1972–1977," from A History of Modern Computing, by Paul E. Ceruzzi, pp. 207–241, text only, © 1998 Massachusetts Institute of Technology, by permission of The MIT Press.

have the disk storage necessary to make it a viable time-sharing system and only about twenty were sold. The PDP-10 did have a random-access disk system, which allowed its users direct access to their own personal files. Like other DEC computers, the PDP-10 also allowed users to load personal files and programs onto inexpensive reels of DECtape, which fitted easily into a briefcase.

The feeling that a PDP-10 was one's own personal computer came from its operating system—especially from the way it managed the flow of information to and from the disks or tapes. With MIT's help, DEC supplied a system called "TOPS-10," beginning in 1972. In the introduction to the TOPS-10 manual, the authors stated, "Our goal has always been that in a properly configured system, each user has the feeling that he owns his portion of the machine for the time he needs to use it." Users could easily create, modify, store, and recall blocks of data from a terminal. The system called these blocks by the already-familiar term, "files." Files were named by one to six characters, followed by a period, then a three-character extension (which typically told what type of file it was, e.g.: xxxxxx.BAS for a program written in BASIC). By typing DIR at a terminal users could obtain a directory of all the files residing on a disk. They could easily send the contents of a file to a desired output device, which typically consisted of a three-letter code, for example, LPT for line printer, or TTY for Teletype.

A small portion of TOPS-10 was always present in core memory. Other programs were stored on the disk and could be called up as necessary. One, called PIP (Peripheral Interchange Program), allowed users to move files in a variety of ways to and from input/output equipment. Another program, TECO (Text Editor and Corrector), allowed users to edit and manipulate text from a terminal. DDT (Dynamic Debugging Tool) allowed users to analyze programs and correct errors without going through the long turnaround times that plagued batch processing.

For PDP-10 users, TOPS-10 was a marvel of simplicity and elegance and gave them the illusion that they were in personal control. TOPS-10 was like a Volkswagen Beetle: basic, simple, and easy to understand and work with. Using a PDP-10 was not only fun but addictive. It was no accident that Brand saw people playing Spacewar on one, or that it was also the computer on which Adventure—perhaps the most long-lasting of all computer games—was written.

On the West Coast another system appeared with similar capabilities, the SDS-940, offered by Scientific Data Systems (SDS) of southern California. The 940 was an extension of a conventional computer, the SDS 930, modified by researchers at Berkeley with support from the Defense Department's Advanced Research Projects Agency. The 940 was more polished than the PDP-10, and it performed well. Still, the PDP-10 seemed to be preferred. At the Xerox Palo Alto Research Center, the legendary lab where so much of personal computing would be created, the staff was encouraged to use SDS machines, since Xerox had just purchased SDS. But the researchers there resisted and instead built a clone of a PDP-10, which they called MAXC—Multiple Access Xerox Computer—the name a pun on Max Palevsky, the founder of SDS. (Palevsky, after becoming very wealthy from the sale of SDS to Xerox, dabbled in Hollywood movies, politics, and culture—and joined the board of *Rolling Stone*. Palevsky also became a venture capitalist with that money, helping to fund Intel, among other companies.)

For a while, when Wall Street was enamored of anything connected with computers, it was easy to raise money to buy or lease a PDP-10 or SDS-940, and then sell computer time to engineering companies or other customers. Most of these firms were undercapitalized and did not understand the complexities of what they were selling. Like their counterparts in the electric utility industry, they had to have enough capacity to handle peak loads, in order not to discourage customers. But that meant that during off-peak times they would be wasting unused and expensive computing equipment. The capital requirements necessary to manage the cycles of the business were as large as they were in the electric power business, which had gone through decades of chaos and turmoil before settling down. Only a few survived, and even fewer, like Tymshare of Cupertino, California, did well (although it was sold to McDonnell-Douglas in the early 1980s). Among those many companies, one is worth mentioning, Computer Center Corporation, or C-Cubed, which installed one of the first PDP-1Os in the Seattle area in 1968. While it was getting started, it offered a local teenager, Bill Gates, free time on the computer in exchange for helping find and rid the system of bugs. C-Cubed folded in 1970, having given Gates a taste of the potential of interactive computing.

Many of those who had access to these systems saw the future of computing. But the financial troubles of time-sharing companies also showed that it would be difficult to make personal, interactive use widely available. There were attempts to make terminals accessible to the public for free or at low cost—the most famous being the Resource One project in the San Francisco Bay area (partially funded by the *Whole Earth Catalog*). But it did not last, either.

■ Calculators and Corporate Personal Computer Projects

Economics prevented the spread of computing to the public from the top down—from large mainframes through time-shared terminals. But while those attempts were underway, the underlying technology was advancing rapidly. Could personal computing arrive from the bottom up—from advances in semiconductor electronics?

Many engineers believe that a mental model of the personal computer was irrelevant. They believe that no one invented the personal computer, it simply flowed from advances in semiconductors. Chuck House, an engineer involved with the early Hewlett-Packard calculators, said, "One could uncharitably say that we invented essentially nothing; we simply took all the ideas that were out there and figured out how to implement them cost-effectively." Gordon Bell stated, "The semiconductor density has really been the driving force, and as you reach different density levels, different machines pop out of that in time." To them, inventions are like a piece of fruit that falls to the ground when it is ripe, and the inventor is given credit for doing little more than picking it up. If that were true, one would find a steady progression of machines offering personal, interactive use, as advances in semiconductors made them viable. And these would have come from established firms who had the engineering and manufacturing resources to translate those advances into products.

Products that took advantage of advances in semiconductors did appear on the market. It is worth looking at them to see whether they validate or refute the bottom-up explanation of the PC's invention.

The first electronic computers were of course operated as if they were personal computers. Once a person was granted access to a machine (after literally waiting in a queue), he or she had the whole computer to use, for whatever purpose. That gave way to more restricted access, but those at MIT and Lincoln Labs who used the Whirlwind, TX-0, and TX-2 that way never forgot its advantages. In 1962 some of them developed a computer called the LINC, made of Digital Equipment Corporation logic modules and intended for use by a researcher as a personal tool. A demonstration project, funded by the NIH, made sixteen LINCs available to biomedical researchers. DEC produced commercial versions, and by the late 1960s, about 1,200 were in use as personal computers. A key feature of the LINC was its compact tape drive and tapes that one could easily carry around: the forerunner of DECtape. The ease of getting at data on the tape was radically different from the clumsy access of tape in mainframes, and this ease would be repeated with the introduction of floppy-disk systems on personal computers. DEC also marketed a computer that was a combination of a LINC and a PDP-8, for $43,000. Although DECtape soon was offered on nearly all DEC's products, the LINC did not achieve the same kind of commercial success as the PDP-8 and PDP-11 lines of minicomputers.

Advances in chip density first made an impact on personal devices in calculators. For decades there had been a small market for machines that could perform the four functions of arithmetic, plus square root. In the 1950s and 1960s the calculator industry was dominated by firms such as Friden and Marchant in the United States, and Odhner in Europe. Their products were complex, heavy, and expensive. In 1964 Wang Laboratories, a company founded by An Wang, a Chinese immigrant who had worked with Howard Aiken at Harvard, came out with an electronic calculator. The Wang LOCI offered more functions, at a lower cost, than the best mechanical machines. Its successor, the Wang 300, was even easier to use and cheaper, partly because Wang deliberately set the price of the 300 to undercut the competitive mechanical calculators from Friden and others. (Only one or two of the mechanical calculator firms survived the transition to electronics.) A few years later Hewlett-Packard, known for its oscilloscopes and electronic test equipment, came out with the HP-9100A, a calculator selling for just under $5,000. And the Italian firm Olivetti came out with the Programma 201, a $3,500 calculator intended primarily for accounting and statistical work. Besides direct calculation, these machines could also execute a short sequence of steps recorded on magnetic cards. Like the LINC, these calculators used discrete circuits. To display digits, the Wang used "Nixie" tubes, an ingenious tube invented by Burroughs in 1957. HP used a small cathode-ray tube, as might be expected from a company that made oscilloscopes.

By 1970 the first of a line of dramatically cheaper and smaller calculators appeared that used integrated circuits. They were about the size of a paperback book and cost as little as $400. A number of wealthy consumers bought them immediately, but it wasn't until Bowmar advertised a Bowmar Brain for less than $250 for the 1971 Christmas season that the calculator burst into public consciousness. Prices plummeted:

under $150 in 1972; under $100 by 1973; under $50 by 1976; finally they became cheap enough to be given away as promotional trinkets. Meanwhile Hewlett-Packard stunned the market in early 1972 with the HP-35, a $400 pocket calculator that performed all the logarithmic and trigonometric functions required by engineers and scientists. Within a few years the slide rule joined the mechanical calculator on the shelves of museums.

Like processed foods, whose cost is mostly in the packaging and marketing, so with calculators: technology no longer determined commercial success. Two Japanese firms with consumer marketing skills, Casio and Sharp, soon dominated. Thirty years after the completion of the half-million dollar ENIAC, digital devices became throw-away commodities. The pioneering calculator companies either stopped making calculators, as did Wang, or went bankrupt, as did Bowmar. Hewlett-Packard survived by concentrating on more advanced and expensive models; Texas Instruments survived by cutting costs.

The commodity prices make it easy to forget that these calculators were ingenious pieces of engineering. Some of them could store sequences of keystrokes in their memory and thus execute short programs. The first of the programmable pocket calculators was Hewlett-Packard's HP-65, introduced in early 1974 for $795. Texas Instruments and others soon followed. As powerful as they were, the trade press was hesitant to call them computers, even if Hewlett-Packard introduced the HP-65 as a "personal computer" (possibly the first use of that term in print). Their limited programming was offset by their built-in ability to compute logarithms and trigonometric functions, and to use floating-point arithmetic to ten decimal digits of precision. Few mainframes could do that without custom-written software.

The introduction of pocket programmable calculators had several profound effects on the direction of computing technology. The first was that the calculator, like the Minuteman and Apollo programs of the 1960s, created a market where suppliers could count on a long production run, and thereby gain economies of scale and a low price. As chip density, and therefore capability, increased, chip manufacturers faced the same problem that Henry Ford had faced with his Model T: only long production runs of the same product led to low prices, but markets did not stay static. That was especially true of integrated circuits, which by nature became ever more specialized in their function as the levels of integration increased. (The only exception was in memory chips, which is one reason why Intel was founded to focus on memories.) The calculator offered the first consumer market for logic chips that allowed companies to amortize the high costs of designing complex integrated circuits. The dramatic drop in prices of calculators between 1971 and 1976 showed just how potent this force was.

The second effect was just as important. Pocket calculators, especially those that were programmable, unleashed the force of personal creativity and energy of masses of individuals. This force had already created the hacker culture at MIT and Stanford (observed with trepidation by at least one MIT professor). Their story is one of the more colorful among the dry technical narratives of hardware and software design. They and their accomplishments, suitably embellished, have become favorite topics of the popular press. Of course their strange personal habits made a good story, but were they true? Developing system software was hard work, not likely to be done well by a salaried employee, working normal hours and with a family to go home to in the evening. Time-sharing freed all users from the tyranny of submitting decks of cards and waiting for a printout, but it forced some users to work late at night, when the time-shared systems were lightly loaded and thus more responsive.

The assertion that hackers created modern interactive computing is about half-right. In sheer numbers there may never have been more than a few hundred people fortunate enough to be allowed to "hack" (that is, not do a programming job specified by one's employer) on a computer like the PDP-10. By 1975, there were over 25,000 HP-65 programmable calculators in use, each one owned by an individual who could do whatever he or she wished to with it. Who were these people? HP-65 users were not "strange". Nearly all were adult professional men, including civil and electrical engineers, lawyers, financial people, pilots, and so on. Only a few were students (or professors), because an HP-65 cost $795. Most purchased the HP-65 because they had a practical need for calculation in their jobs. But this was a *personal* machine—one could take it home at night. These users—perhaps 5 or 10 percent of those who owned machines—did not fit the popular notion of hackers as kids with "[t]heir rumpled clothes, their unwashed and unshaven faces, and their uncombed hair." But their passion for programming made them the intellectual cousins of the students in the Tech Model Railroad Club. And their numbers—only to increase as the prices of calculators dropped—were the first indication that personal computing was truly a mass phenomenon.

Hewlett-Packard and Texas Instruments were unprepared for these events. They sold the machines as commodities; they could ill-afford a sales force that could walk a customer through the complex learning process needed to get the most out of one. That was what IBM salesmen were known for—but they sold multimillion dollar mainframes. Calculators were designed to be easy enough to use to make that unnecessary, at least for basic tasks. What was unexpected was how much more some of those customers wanted to do. Finding little help from the supplier, they turned to one another. Users groups, clubs, newsletters, and publications proliferated.

This supporting infrastructure was critical to the success of personal computing; in the following decade it would become an industry all its own. Many histories of the personal computer emphasize this point; they often cite the role of the Homebrew Computer Club, which met near the Stanford campus in the mid-1970s, as especially important. The calculator users groups were also important, though for different reasons. As the primitive first personal computers like the Altair gave way to more complete systems, a number of calculator owners purchased one of them as well. In the club newsletters there were continuous discussions of the advantages and drawbacks of each—the one machine having the ability to evaluate complex mathematical expressions with ease, the other more primitive but *potentially* capable of doing all that and more. There was no such thing as a typical member of the Homebrew Computer Club, although calculator owners tended to be professionals whose jobs required calculation during the day, and who thought of other uses at night. Many of them were bitten by the PC bug; at the same time they took a show-me attitude toward the computer. Could you rely on one? Could you use one to design a radar antenna? Could it handle a medium-sized mailing list? Was the personal computer a serious machine? At first the answers were, "not yet," but gradually, with some firm prodding by this community, the balance shifted. Groups like the Homebrew Computer Club emphasized the "personal" in personal computer; calculator users emphasized the word computer.

Ever since time-sharing and minicomputers revealed an alternative to mainframe computing, there have been prophets and evangelists who raged against the world of punched cards and computer rooms, promising a digital paradise of truly interactive tools. The most famous was Ted Nelson, whose self-published book *Computer Lib* proclaimed (with a raised fist on the cover): "You can and must understand computers *now*." By 1974 enough of these dreams had become real that the specific abilities—and limits—of actual "dream machines" (the alternate title to Nelson's book) had to be faced. Some of the dreamers, including Nelson, were unable to make the transition. They dismissed the pocket calculator. They thought it was puny, too cheap, couldn't do graphics, wasn't a "von Neumann machine," and so on. For them, the dream machine was better, even if (or because) it was unbuilt. By 1985 there would be millions of IBM Personal Computers and their copies in the offices and homes of ordinary people. These computers would use a processor that was developed for other purposes, and adapted for the personal computer almost by accident. But they would be real and a constant source of inspiration and creativity to many who used them, as well as an equal source of frustration for those who knew how much better they could be.

■ The Microprocessor

Calculators showed what integrated circuits could do, but they did not open up a direct avenue to personal interactive computing. The chips used in them were too specialized for numerical calculation to form a basis for a general-purpose computer. Their architecture was ad-hoc and closely guarded by each manufacturer. What was needed was a set of integrated circuits—or even a single integrated circuit—that incorporated the basic architecture of a general-purpose, stored-program computer. Such a chip, called a "microprocessor," did appear.

In 1964 Gordon Moore, then of Fairchild and soon a cofounder of Intel, noted that from the time of the invention of integrated circuits in 1958, the number of circuits that one could place on a single integrated circuit was doubling every year. By simply plotting this rate on a piece of semi-log graph paper, "Moore's Law" predicted that by the mid 1970s one could buy a chip containing logic circuits equivalent to those used in a 1950s-era mainframe. (Recall that the UNIVAC I had about 3,000 tubes, about the same number of active elements contained in the first microprocessor discussed below.) By the late 1960s transistor-transistor logic (TTL) was well established, but a new type of semiconductor called metal-oxide semiconductor (MOS), emerged as a way to place even more logic elements on a chip. MOS was used by Intel to produce its pioneering

1103 memory chip, and it was a key to the success of pocket calculators. The chip density permitted by MOS brought the concept of a computer-on-a-chip into focus among engineers at Intel, Texas Instruments, and other semiconductor firms. That did not mean that such a device was perceived as useful. If it was generally known that enough transistors could be placed on a chip to make a computer, it was also generally believed that the market for such a chip was so low that its sales would never recoup the large development costs required.

By 1971 the idea was realized in silicon. Several engineers deserve credit for the invention. Ted Hoff, an engineer at Intel, was responsible for the initial concept, Federico Faggin of Intel deserves credit for its realization in silicon, and Gary Boone of Texas Instruments designed similar circuits around that time. In 1990, years after the microprocessor became a household commodity and after years of litigation, Gil Hyatt, an independent inventor from La Palma, California, received a patent on it. Outside the courts he has few supporters, and recent court rulings may have invalidated his claim entirely.

The story of the microprocessor's invention at Intel has been told many times. In essence, it is a story encountered before: Intel was asked to design a special-purpose system for a customer. It found that by designing a general-purpose computer and using software to tailor it to the customer's needs, the product would have a larger market.

Intel's customer for this circuit was Busicom, a Japanese company that was a top seller of hand-held calculators. Busicom sought to produce a line of products with different capabilities, each aimed at a different market segment. It envisioned a set of custom-designed chips that incorporated the logic for the advanced mathematical functions. Intel's management assigned Marcian E. ("Ted") Hoff, who had joined the company in 1968 (Intel's twelfth employee), to work with Busicom.

Intel's focus had always been on semiconductor memory chips. It had shied away from logic chips like those suggested by Busicom, since it felt that markets for them were limited. Hoff's insight was to recognize that by designing fewer logic chips with more general capabilities, one could satisfy Busicom's needs elegantly. Hoff was inspired by the PDP-8, which had a very small set of instructions, but which its thousands of users had programmed to do a variety of things. He also recalled using an IBM 1620, a small scientific computer with an extremely limited instruction set that nevertheless could be programmed to do a lot of useful work.

Hoff proposed a logic chip that incorporated more of the concepts of a general-purpose computer. A critical feature was the ability to call up a subroutine, execute it, and return to the main program as needed. He proposed to do that with a register that kept track of where a program was in its execution and saved that status when interrupted to perform a subroutine. Subroutines themselves could be interrupted, with return addresses stored on a "stack": an arrangement of registers that automatically retrieved data on a last-in-first-out basis.

With this ability, the chip could carry out complex operations stored as subroutines in memory, and avoid having those functions permanently wired onto the chip. Doing it Hoff's way would be slower, but in a calculator that did not matter, since a person could not press keys that fast anyway. The complexity of the logic would now reside in software stored in the memory chips, so one was not getting something for nothing. But Intel was a memory company, and it knew that it could provide memory chips with enough capacity. As an added inducement, sales of the logic chips would mean more sales of its bread-and-butter memories.

That flexibility meant that the set of chips could be used for many other applications besides calculators. Busicom was in a highly competitive and volatile market, and Intel recognized that. (Busicom eventually went bankrupt.) Robert Noyce negotiated a deal with Busicom to provide it with chips at a lower cost, giving Intel in return the right to market the chips to other customers for noncalculator applications. From these unsophisticated negotiations with Busicom, in Noyce's words, came a pivotal moment in the history of computing.

The result was a set of four chips, first advertised in a trade journal in late 1971, which included "a microprogrammable computer on a chip!" That was the 4004, on which one found all the basic registers and control functions of a tiny, general-purpose stored-program computer. The other chips contained a read-only memory (ROM), random-access memory (RAM), and a chip to handle output functions. The 4004 became the historical milestone, but the other chips were important as well, especially the ROM chip that supplied the code that turned a general-purpose processor into something that could meet a customer's needs. (Also at Intel, a team led by Dov Frohman developed a ROM chip that could be easily reprogrammed and erased by exposure to ultraviolet light. Called an EPROM (erasable programmable read-only memory) and introduced in 1971, it made the concept of system design using a microprocessor practical.)

The detailed design of the 4004 was done by Stan Mazor. Federico Faggin was also crucial in making the concept practical. Masatoshi Shima, a representative from Busicom, also contributed. Many histories of the invention give Hoff sole credit; all players, including Hoff, now agree that that is not accurate. Faggin left Intel in 1974 to found a rival company, Zilog. Intel, in competition with Zilog, felt no need to advertise Faggin's talents in its promotional literature, although Intel never showed any outward hostility to its ex-employee. The issue of whom to credit reveals the way many people think of invention: Hoff had the idea of putting a general-purpose computer on a chip. Faggin and the others "merely" implemented that idea in silicon. At the time, Intel was not sure what it had invented either: Intel's patent attorney resisted Hoff's desire at the time to patent the work as a "computer." Intel obtained two patents on the 4004, covering its architecture and implementation; Hoff's name appears on only one of them. (That opened the door to rival claims for patent royalties from TI, and eventually Gil Hyatt.)

The 4004 worked with groups of four bits at a time—enough to code decimal digits but no more. At almost the same time as the work with Busicom, Intel entered into a similar agreement with Computer Terminal Corporation (later called Datapoint) of San Antonio, Texas, to produce a set of chips for a terminal to be attached to mainframe computers. Again, Mazor and Hoff proposed a microprocessor to handle the terminal's logic. Their proposed chip would handle data in 8-bit chunks, enough to process a full byte at a time. By the time Intel had completed its design, Datapoint had decided to go with conventional TTL chips. Intel offered the chip, which they called the 8008, as a commercial product in April 1972.

In late 1972, a 4-bit microprocessor was offered by Rockwell, an automotive company that had merged with North American Aviation, maker of the Minuteman Guidance System. In 1973 a half dozen other companies began offering microprocessors as well. Intel responded to the competition in April 1974 by announcing the 8080, an 8-bit chip that could address much more memory and required fewer support chips than the 8008. The company set the price at $360—a somewhat arbitrary figure, as Intel had no experience selling chips like these one at a time. (Folklore has it that tire $360 price was set to suggest a comparison with the IBM System/360.) A significant advance over the 8008, the 8080 could execute programs written for the other chip, a compatibility that would prove crucial to Intel's dominance of the market. The 8080 was the first of the microprocessors whose instruction set and memory addressing capability approached those of the mini-computers of the day.

■ From Microprocessor to Personal Computer

There were now, in early 1974, two converging forces at work. From one direction were the semiconductor engineers with their ever-more-powerful microprocessors and ever-more-capacious memory chips. From the other direction were users of time-sharing systems, who saw a PDP-10 or XDS 940 as a basis for public access to computing. When these forces met in the middle, they would bring about a revolution in personal computing.

They almost did not meet. For the two years between Brand's observation and the appearance of tire Altair, the two forces were rushing past one another. The time-sharing systems had trouble making money even from industrial clients, and the public systems like community Memory were also struggling. At the other end, semiconductor companies did not think of their products as a possible basis for a personal computer.

A general-purpose computer based on a microprocessor did appear in 1973. In May of that year Thi T. Truong, an immigrant to France from Viet Nam, had his electronics company design and build a computer based on the Intel 8008 microprocessor. The MICRAL was a rugged and well-designed computer, with a bus architecture and internal slots on its circuit board for expansion. A base model cost under $2,000, and it found a market replacing minicomputers for simple control operations. Around two thousand were sold in the next two years, none of them beyond an industrial market. It is regarded as the first microprocessor-based computer to be sold in the commercial marketplace. Because of the limitations of the 8008, its location in France, and above all, the failure by its creators to see what it "really" was, it never broke out of its niche as a replacement for minicomputers in limited industrial locations.

The perception of the MICRAL as something to replace the mini was echoed at Intel as well. Intel's mental model of its product was this: an *industrial* customer bought an 8080 and wrote specialized software for it, which was then burned into a read-only-memory to give a system with the desired functions. The resulting inexpensive product (no longer programmable) was then put on the market as an embedded controller in an industrial system. A major reason for that mental model was the understanding of how hard it was to program a microprocessor. It seemed absurd to ask untrained consumers to program when Intel's traditional customers, hardware designers, were themselves uncomfortable with programming.

With these embedded uses in mind, microprocessor suppliers developed educational packages intended to ease customers into system design. These kits included the microprocessor, some RAM and ROM chips, and some other chips that handled timing and control, all mounted on a printed circuit board. They also included written material that gave a tutorial on how to program the system. This effort took Intel far from its core business of making chips, but the company hoped to recoup the current losses later on with volume sales of components. These kits were sold for around $200 or given away to engineers who might later generate volume sales.

Intel and the others also built more sophisticated "Development Systems," on which a customer could actually test the software for an application. These were fully assembled products that sold for around $10,000. To use these systems, customers also needed specialized software that would allow them to write programs using a language like FORTRAN, and then "cross-compile" it for the microprocessor—that is, from the FORTRAN program generate machine code, not for the computer on which it was written, but for the microprocessor. The company hired Gary Kildall, an instructor at the Naval Postgraduate School in Monterey, California, to develop a language based on IBM's PL/I. He called it PL/M, and in 1973 Intel offered it to customers. Initially this software was intended to be run on a large mainframe, but it was soon available for minicomputers, and finally to microprocessor-based systems. In 1974 Intel offered a development system, the Intellec 4, which included its own resident PL/M compiler (i.e., one did not need a mainframe or a mini to compile the code). A similar Intellec-8 introduced the 8-bit microprocessors.

With these development systems, Intel had in fact invented a personal computer. But the company did not realize it. These kits were not marketed as the functional computers they were. Occasionally someone bought one of these systems and used it in place of a minicomputer, but Intel neither supported that effort nor recognized its potential. Intel and the other microprocessor firms made money selling these development systems—for some they were very profitable—but the goal was to use them as a lever to open up volume purchases of chips. The public could not buy one. The chip suppliers were focused on the difficulties in getting embedded systems to do useful work; they did not think that the public would be willing to put up with the difficulties of programming just to own their own computer.

▪ Role of Hobbyists

Here is where the electronics hobbyists and enthusiasts come in. Were it not for them, the two forces in personal computing might have crossed without converging. Hobbyists, at that moment, were willing to do the work needed to make microprocessor-based systems practical.

This community had a long history of technical innovation—it was radio amateurs, for example, who opened up the high-frequency radio spectrum for long-distance radio communications after World War I. After World War II, the hobby expanded beyond amateur radio to include high-fidelity music reproduction, automatic controls, and simple robotics. A cornucopia of war surplus equipment from the U.S. Army Signal Corps found its way into individual hands, further fueling the phenomenon. (A block in lower Manhattan known as "Radio Row," where the World Trade Center was built, was a famous source of surplus electronic gear.) The shift from vacuum tubes to integrated circuits made it harder for an individual to build a circuit on a breadboard at home, but inexpensive TTL chips now contained whole circuits themselves. As the hobby evolved rapidly from analog to digital applications, this group supplied a key component in creating the personal computer: it provided an infrastructure of support that neither the computer companies nor the chip makers could.

This infrastructure included a variety of electronics magazines. Some were aimed at particular segments, for example, *QST* for radio amateurs. Two of them, *Popular Electronics* and *Radio-Electronics*, were of general interest and sold at newsstands; they covered high-fidelity audio, shortwave radio, television, and assorted gadgets for the home and car. Each issue typically had at least one construction project. For these projects the magazine would make arrangements with small electronics companies to supply a printed circuit board, already etched and drilled, as well as specialized components that readers might have difficulty finding locally. By scanning the back issues of these magazines we can trace how hobbyists moved from analog to digital designs.

A machine called the Kenbak-1, made of medium and small-scale integrated circuits, was advertised in the September 1971 issue of *Scientific American*. The advertisement called it suitable for "private individuals," but it was really intended for schools. The Kenbak may be the first personal computer, but it did not use a microprocessor, and its capabilities were quite limited.

The Scelbi-8H was announced in a tiny advertisement in the back of the March 1974 issue of *QST*. It used an Intel 8008, and thus may be the first microprocessor-based computer marketed to the public. According to the advertisement, "Kit prices for the new Scelbi-8H mini-computer start as low as $440!" It is not known how many machines Scelbi sold, but the company went on to play an important part in the early personal computer phenomenon.

In July 1974, *Radio-Electronics* announced a kit based on the Intel 8008, under the headline "Build the Mark-8: Your Personal Minicomputer." The project was much more ambitious than what typically appeared in that magazine. The article gave only a simple description and asked readers to order a separate, $5.00 booklet for detailed instructions. The Mark-8 was designed by Jonathan Titus of Virginia Polytechnic University in Blacksburg. The number of machines actually built may range in the hundreds, although the magazine reportedly sold "thousands" of booklets. At least one Mark-8 users club sprang up, in Denver, whose members designed an ingenious method of storing programs on an audio cassette recorder. Readers were directed to a company in Englewood, New Jersey, that supplied a set of circuit boards for $47.50, and to Intel for the 8008 chip (for $120.00). The Mark-8's appearance in *Radio-Electronics* was a strong factor in the decision by its rival *Popular Electronics* to introduce the Altair kit six months later.

These kits were just a few of many projects described in the hobbyist magazines. They reflected a conscious effort by the community to bring digital electronics, with all its promise and complexity, to amateurs who were familiar only with simpler radio or audio equipment. It was not an easy transition: construction of both the Mark-8 and the TV-typewriter (described next) was too complex to be described in a magazine article; readers had to order a separate booklet to get complete plans. *Radio-Electronics* explained to its readers that "[w]e do not intend to do an article this way as a regular practice." Although digital circuits were more complex than what the magazine had been handling, it recognized that the electronics world was moving in that direction and that its readers wanted such projects.

Other articles described simpler digital devices—timers, games clocks, keyboards, and measuring instruments—that used inexpensive TTL chips. One influential project was the TV-Typewriter, designed by Don Lancaster and published in *Radio-Electronics* in September 1973. This device allowed readers to display alphanumeric characters, encoded in ASCII, on an ordinary television set. It presaged the advent of CRT terminals as the primary input-output device for personal computers—one major distinction between the PC culture and that of the minicomputer, which relied on the Teletype. Lee Felsenstein called the TV-Typewriter "the opening shot of the computer revolution."

■ Altair

1974 was the *anmus mirabilis* of personal computing. In January, Hewlett-Packard introduced its HP-65 programmable calculator. That summer Intel announced the 8080 microprocessor. In July, *Radio-Electronics* described the Mark-8. In late December, subscribers to *Popular Electronics* received their January 1975 issue in the mail, with a prototype of the "Altair" minicomputer on the cover, and an article describing how readers could obtain one for less than $400. This announcement ranks with IBM's announcement of the System/360 a decade earlier as one of the most significant in the history of computing. But what a difference a decade made: the Altair was a genuine personal computer.

H. Edward Roberts, the Altair's designer, deserves credit as the inventor of the personal computer. The Altair was a capable, inexpensive computer designed around the Intel 8080 microprocessor. Although calling Roberts the inventor makes sense only in the context of all that came before him, including the crucial steps described above, he does deserve the credit. Mark Twain said that historians have to rearrange past events so they make more sense. If so, the invention of the personal computer at a small model-rocket hobby shop in

Albuquerque cries out for some creative rearrangement. Its utter improbability and unpredictability have led some to credit many other places with the invention, places that are more sensible, such as the Xerox Palo Alto Research Center, or Digital Equipment Corporation, or even IBM. But Albuquerque it was, for it was only at MITS that the technical and social components of personal computing converged.

Consider first the technical. None of the other hobbyist projects had the impact of the Altair's announcement. Why? One reason was that it was designed and promoted as a capable minicomputer, as powerful as those offered by DEC or Data General. The magazine article, written by Ed Roberts and William Yates, makes this point over and over: "a full-blown computer that can hold its own against sophisticated minicomputers"; "not a 'demonstrator' or a souped-up calculator"; "performance competes with current commercial minicomputers." The physical appearance of the Altair computer suggested its minicomputer lineage. It looked like the Data General Nova: it had a rectangular metal case, a front panel of switches that controlled the contents of internal registers, and small lights indicating the presence of a binary one or zero. Inside the Altair's case, there was a machine built mainly of TTL integrated circuits (except for the microprocessor, which was a MOS device), packaged in dual-in-line packages, soldered onto circuit boards. Signals and power traveled from one part of the machine to another on a bus. The Altair used integrated circuits, not magnetic cores, for its primary memory. The *Popular Electronics* cover called the Altair the "world's first minicomputer kit"; except for its use of a microprocessor, that accurately described its physical construction and design.

But the Altair as advertised was ten times cheaper than minicomputers were in 1975. The magazine offered an Altair for under $400 as a kit, and a few hundred more already assembled. The magazine cover said that readers could "save over $1,000." In fact, the cheapest PDP-8 cost several thousand dollars. Of course, a PDP-8 was a fully assembled, operating computer that was considerably more capable than the basic Altair, but that did not really matter in this case. (Just what one got for $400 will be discussed later.) The low cost resulted mainly from its use of the Intel 8080 microprocessor, just introduced. Intel had quoted a price of $360 for small quantities of 8080s, but Intel's quote was not based on a careful analysis of how to sell the 8080 to this market. MITS bought them for only $75 each.

The 8080 had more instructions and was faster and more capable than the 8008 that the Mark-8 and Scelbi-8 used. It also permitted a simpler design since it required only six instead of twenty supporting chips to make a functional system. Other improvements over the 8008 were its ability to address up to 64 thousand bytes of memory (vs. the 8008's 16 thousand), and its use of main memory for the stack, which permitted essentially unlimited levels of subroutines instead of the 8008's seven levels.

The 8080 processor was only one architectural advantage the Altair had over its predecessors. Just as important was its use of an open bus. According to folklore, the bus architecture almost did not happen. After building the prototype Altair, Roberts photographed it and shipped it via Railway Express to the offices of *Popular Electronics* in New York. Railway Express, a vestige of an earlier American industrial revolution, was about to go bankrupt; it lost the package. The magazine cover issue showed the prototype, with its light-colored front panel and the words "Altair 8800" on the upper left. That machine had a set of four large circuit boards stacked on top of one another, with a wide ribbon cable carrying 100 lines from one board to another. After that machine was lost, Robert redesigned the Altair. He switched to a larger deep blue cabinet and discarded the 100-wire ribbon cable. In the new design, wires connected to a rigid backplane carried the signals from one board to another. That allowed hobbyists to add a set of connectors that could accept other cards besides the initial four.

The $400 kit came with only two cards to plug into the bus: those two, plus a circuit board to control the front panel and the power supply, made up the whole computer. The inside looked quite bare. But laboriously soldering a set of wires to an expansion chassis created a full set of slots into which a lot of cards could be plugged. MITS was already designing cards for more memory, I/O and other functions.

Following the tradition established by Digital Equipment Corporation, Roberts did not hold specifications of the bus as a company secret. That allowed others to design and market cards for the Altair. That decision was as important to the Altair's success as its choice of ar 8080 processor. It also explains one of the great ironies of the Altair, that it inaugurated the PC era although it was neither reliable nor very well-designed. Had it not been possible for other companies to offer plug-in cards that improved on the original MITS design, the Altair might have made no greater impact than the Mark-8 had. The bus architecture also led to the company's demise a few years later, since it allowed other companies to market compatible cards and, later, compatible computers. But by then the floodgates had opened. If MITS was unable lo deliver on its promises of making the Altair a serious machine (though it tried), other companies would step in. MITS continued developing plug-in cards and peripheral equipment, but the flood of orders was too much for the small company.

So while it was true that for $400 hobbyists got very little, they could get the rest—or design and build the rest. Marketing the computer as a bare-bones kit offered a way for thousands of people to bootstrap their way into the computer age, at a pace that they, not a computer company, could control.

Assembling the Altair was much more difficult than assembling other electronics kits, such as those sold by the Heath Company or Dynaco. MITS offered to sell "completely assembled and tested" computers for $498, but with such a backlog of orders, readers were faced with the choice of ordering the kit and getting something in a couple of months, or ordering the assembled computer and perhaps waiting a year or more. Most ordered the kit and looked to one another for support in finding the inevitable wiring errors and poorly soldered connections that they would make. The audience of electronics hobbyists, at whom the magazine article was aimed, compared the Altair not to the simple Heathkits, but to building a computer from scratch, which was almost impossible: not only was it hard to design a computer, it was impossible to obtain the necessary chips. Chips were inexpensive, but only if they were purchased in large quantities, and anyway, most semiconductor firms had no distribution channels set up for single unit or retail sales. Partly because of this, customers felt, rightly, that they were getting an incredible bargain.

The limited capabilities of the basic Altair, plus the loss of the only existing Altair by the time the *Popular Electronics* article appeared, led to the notion that it was a sham, a "humbug," not a serious product at all. The creators of the Altair fully intended to deliver a serious computer whose capabilities were on a par with minicomputers then on the market. Making those deliveries proved to be a lot harder than they anticipated. Fortunately, hobbyists understood that. But there should be no mistake about it: the Altair was real.

MITS and the editors of *Popular Electronics* had found a way to bring the dramatic advances in integrated circuits to individuals. The first customers were hobbyists, and the first thing they did with these machines, once they got them running, was play games. Roberts was trying to sell it as a machine for serious work, however. In the *Popular Electronics* article he proposed a list of twenty-three applications, none of them games. Because it was several years before anyone could supply peripheral equipment, memory, and software, serious applications were rare at first. That, combined with the primitive capabilities of other machines like the Mark-8, led again to an assumption that the Altair was not a serious computer. Many of the proposed applications hinted at in the 1975 article were eventually implemented. Years later one could still find an occasional Altair (or more frequently, an Altair clone) embedded into a system just like its minicomputer cousins.

The next three years, from January 1975 through the end of 1977, saw a burst of energy and creativity in computing that had almost no equal in its history. The Altair had opened the floodgates, even though its shortcomings were clear to everyone. One could do little more than get it to blink a pattern of lights on the front panel. And even that was not easy: one had to flick the toggle switches for each program step, then deposit that number into a memory location, then repeat that for the next step, and so on—hopefully the power did not go off while this was going on—until the whole program (less than 256 bytes long!) was in memory. Bruised fingers from flipping the small toggle switches were the least of the frustrations. In spite of all that, the bus architecture meant that other companies could design boards to remedy each of these Shortcomings, or even design a copy of the Altair itself, as IMSAI and others did.

But the people at MITS and their hangers-on created more than just a computer. This $400 computer inspired the extensive support of user groups, informal newsletters, commercial magazines, local clubs, conventions, and even retail stores. This social activity went far beyond traditional computer user groups, like SHARE for IBM or DECUS for Digital. Like the calculator users groups, these were open and informal, and offered more to the neophyte. All of this sprang up with the Altair, and many of the publications and groups lived long after the last Altair computer itself was sold.

Other companies, beginning with Processor Technology, soon began offering plug-in boards that gave the machine more memory. Another board provided a way of connecting the machine to a Teletype, which allowed fingers to heal. But Teletypes were not easy to come by—an individual not affiliated with a corporation or university could only buy one secondhand, and even then they were expensive. Before long, hobbyists-led small companies began offering ways of hooking up a television set and a keyboard (although Don Lancaster's

TV Typewriter was not the design these followed). The board that connected to the Teletype sent data serially—one bit at a time; another board was designed that sent out data in parallel, for connection to a line printer that minicomputers used, although like the Teletype these were expensive and hard to come by.

The Altair lost its data when the power was shut off, but before long MITS designed an interface that put out data as audio tones, to store programs on cheap audio cassettes. A group of hobbyists met in Kansas City in late 1975 and established a "Kansas City Standard" for the audio tones stored on cassettes, so that programs could be exchanged from one computer to another. Some companies brought out inexpensive paper tape readers that did not require the purchase of a Teletype. Others developed a tape cartridge like the old 8-track audio systems, which looped a piece of tape around and around. Cassette storage was slow and cumbersome—users usually had to record several copies of a program and make several tries before successfully loading it into the computer. Inadequate mass storage limited the spread of PCs until the "floppy" disk was adapted.

The floppy was invented by David L. Noble at IBM for a completely different purpose. When IBM introduced the System/370, which used semiconductor memory, it needed a way to store the computer's initial control program, as well as to hold the machine's microprogram. That had not been a problem for the System/360, which used magnetic cores that held their contents when the power was switched off. From this need came the 8-inch diameter flexible diskette, which IBM announced in 1971. Before long, people recognized that it could be used for other purposes besides the somewhat limited one for which it had been invented. In particular, Alan Shugart, who had once worked for IBM, recognized that the floppy's simplicity and low cost made it the ideal storage medium for low-cost computer systems. Nevertheless, floppy drives were rare in the first few years of personal computing. IBM's hardware innovation was not enough; there had to be an equivalent innovation in system software to make the floppy practical. Before that story is told, we shall look first at the more immediate issue of developing a high-level language for the PC.

■ Software: BASIC

The lack of a practical mass storage device was one of two barriers that blocked the spread of personal, interactive computing. The other was a way to write applications software. By 1977 two remarkable and influential pieces of software—Microsoft BASIC and the CP/M Operating System—overcame those barriers.

In creating the Altair, Ed Roberts had to make a number of choices: what processor to use, the design of the bus (even whether to use a bus at all), the packaging, and so on. One such decision was the choice of a programming language. Given the wide acceptance of BASIC it is hard to imagine that there ever was a choice, but there was. BASIC was not invented for small computers. The creators of BASIC abhorred the changes others made to shoehorn the language onto systems smaller than a mainframe. Even in its mainframe version, BASIC had severe limitations—on the numbers and types of variables it allowed, for example. In the view of academic computer scientists, the versions of BASIC developed for minicomputers were even worse—full of ad hoc patches and modifications. Many professors disparaged BASIC as a toy language that fostered poor programming habits, and they refused to teach it. Serious programming was done in FORTRAN—an old and venerable but still capable language.

If, in 1974, one asked for a modern, concise, well-designed language to replace FORTRAN, the answer might have been APL, an interactive language invented at IBM by Kenneth Iverson in the early 1960s. A team within IBM designed a personal computer in 1973 that supported APL, the "SCAMP," although a commercial version of that computer sold poorly. Or PL/I: IBM had thrown its resources into this language, which it hoped would replace both FORTRAN and COBOL. Gary Kildall chose a subset of PL/I for the Intel microprocessor development kit.

BASIC'S strength was that it was easy to learn. More significant, it already had a track record of running on computers with limited memory. Roberts stated that he had considered FORTRAN and APL, before he decided the Altair was to have BASIC.

William Gates III was born in 1955, at a time when work on FORTRAN was just underway. He was a student at Harvard when the famous cover of *Popular Electronics* appeared describing the Altair. According to one biographer, his friend Paul Allen saw the magazine and showed it to Gates, and the two immediately decided that they would write a BASIC compiler for the machine. Whether it was Gates's or Roberts's decision to go with BASIC for the Altair, BASIC it was.

In a newsletter sent out to Altair customers, Gates and Allen stated that a version of BASIC that required only 4K bytes of memory would be available in June 1975, and that more powerful versions would be available soon after. The cost, for those who also purchased Altair memory boards, was $60 for 4K BASIC, $75 for 8K, and $150 for "extended" BASIC (requiring disk or other mass storage). Those who wanted the language to run on another 8080-based system had to pay $500.

In a burst of energy, Gates and Allen, with the help of Monte Davidoff, wrote not only a BASIC that fit into very little memory; they wrote a BASIC with a lot of features and impressive performance. The language was true to its Dartmouth roots in that it was easy to learn. It broke with those roots by providing a way to move from BASIC commands to instructions written in machine language. That was primarily through a USR command, which was borrowed from software written for DEC minicomputers (where the acronym stood for user service routine). A programmer could even directly put bytes into or pull data out of specific memory locations, through the PEEK and POKE commands—which would have caused havoc on the time-shared Dartmouth system. Like USR, these commands were also derived from prior work done by DEC programmers, who came up with them for a time-sharing system they wrote in BASIC for the PDP-11. Those commands allowed users to pass from BASIC to machine language easily—a crucial feature for getting a small system to do useful work.

These extensions kept their BASIC within its memory constraints while giving it the performance of a more sophisticated language. Yet it remained an interactive, conversational language that novices could learn and use. The BASIC they wrote for the Altair, with its skillful combination of features taken from Dartmouth and from the Digital Equipment Corporation, was the key to Gates's and Allen's success in establishing a personal computer software industry.

The developers of this language were not formally trained in computer science or mathematics as were Kemeny and Kurtz. They were introduced to computing in a somewhat different way. Bill Gates's private school in Seattle had a General Electric time-sharing system available for its pupils in 1968, a time when few students even in universities had such access. Later on he had access to an even better time-shared system: a PDP-10 owned by the Computer Center Corporation. Later still, he worked with a system of PDP-10s and PDP-11s used to control hydroelectric power for the Bonneville Power Administration. One of his mentors at Bonneville Power was John Norton, a TRW employee who had worked on the Apollo Program and who was a legend among programmers for the quality of his work.

When he was writing BASIC for the Altair, Gates was at Harvard. He did not have access to an 8080-based system, but he did have access to a PDP-10 at Harvard's computing center (named after Howard Aiken). He and fellow student Monte Davidoff used the PDP-10 to write the language, based on the written specifications of the Intel 8080. In early 1975 Paul Allen flew to Albuquerque and demonstrated it to Roberts and Yates. It worked. Soon after, MITS advertised its availability for the Altair. Others were also writing BASIC interpreters for the Altair and for the other small computers now flooding the market, but none was as good as Gates's and Allen's, and it was not long before word of that got around.

It seemed that Roberts and his company had made one brilliant decision after another: the 8080 processor, the bus architecture, and now BASIC. However, by late 1975 Gates and Allen were not seeing it that way. Gates insists that he never became a MITS employee (although Allen was until 1976), and that under the name "Micro Soft," later "Micro-Soft," he and Allen retained the rights to their BASIC. In a now-legendary "Open Letter to Hobbyists," distributed in early 1976, Gates complained about people making illicit copies of his BASIC by duplicating the paper tape. Gates claimed "the value of the computer time we have used [to develop the language] exceeds $40,000." He said that if he and his programmers were not paid, they would have little incentive to develop more software for personal computers, such as an APL language for the 8080 processor. He argued that illicit copying put all personal computing at risk: "Nothing would please me more than to hire ten programmers and deluge the hobby market with good software."

Gates did his initial work on the PDP-10 while still an undergraduate at Harvard. Students were not to use that computer for commercial purposes, although these distinctions were not as clear then as they would be later. The language itself was the invention of Kemeney and Kurtz of Dartmouth; the extensions that were crucial to its success came from programmers at the Digital Equipment Corporation, especially Mark Bramhall, who led the effort to develop a time-sharing system (RSTS-11) for the PDP-11. Digital, the only commercial entity among the above group, did not think of its software as a commodity to sell; it was what the company did to get people to buy hardware.

Bill Gates had recognized what Roberts and all the others had not: that with the advent of cheap, personal computers, software could and should come to the fore as the principal driving agent in computing. And only by charging money for it—even though it had originally been free—could that happen. By 1978 his company, now called "Microsoft," had severed its relationship with MITS and was moving from Albuquerque to the Seattle suburb of Bellevue. (MITS itself had lost its identity, having been bought by Pertec in 1977.) Computers were indeed coming to "the people," as Stewart Brand had predicted in 1972. But the driving force was not the counterculture vision of a Utopia of shared and free information; it was the force of the market-place. Gates made good on his promise to "hire ten programmers and deluge the . . . market".

■ System Software: The Final Piece of the Puzzle

Gary Kildall's entree into personal computing software was as a consultant for Intel, where he developed languages for system development. While doing that he recognized that the floppy disk would make a good mass storage device for small systems, if it could be properly adapted. To do that he wrote a small program that managed the flow of information to and from a floppy disk drive. As with the selection of BASIC, it appears in hindsight to be obvious and inevitable that the floppy disk would be the personal computer's mass storage medium. That ignores the fact that it was never intended for that use. As with the adaptation of BASIC, the floppy had to be recast into a new role. As with BASIC, doing that took the work of a number of individuals, but the primary effort came from one man, Gary Kildall.

A disk had several advantages over magnetic or paper tape. For one, it was faster. For another, users could both read and write data on it. Its primary advantage was that a disk had "random" access: Users did not have to run through the entire spool of tape to get at a specific piece of data. To accomplish this, however, required tricky programming—something IBM had called, for one of its mainframe systems, a Disk Operating System, or DOS.

A personal computer DOS had little to do with mainframe operating systems. There was no need to schedule and coordinate the jobs of many users: an Altair had one user. There was no need to "spool" or otherwise direct data to a roomful of chain printers, card punches, and tape drives: a personal computer had only a couple of ports to worry about. What *was* needed was rapid and accurate storage and retrieval of files from a floppy disk. A typical file would, in fact, be stored as a set of fragments, inserted at whatever free spaces were available on the disk. It was the job of the operating system to find those free spaces, put data there, retrieve it later on, and reassemble the fragments. All that gave the user an illusion that the disk was just like a traditional file cabinet filled with folders containing paper files.

Once again, Digital Equipment Corporation was tire pioneer, in part because of its culture; because of the experience many of its employees had had with the TX-0 at MIT, one of the first computers to have a conversational, interactive feel to it. For its early systems DEC introduced DECtape, which although a tape, allowed programmers rapid access to data written in the middle, as well as at the ends, of the reel. The PDP-10s had powerful DECtape as well as disk storage abilities; its operating systems were crucial in creating the illusion of personal computing that had so impressed observers like Stewart Brand.

In the late 1960s DEC produced OS/8 for the PDP-8, which had the feel of the PDP-10 but ran on a machine with very limited memory. OS-8 opened everyone's eyes at DEC; it showed that small computers could have capabilities as sophisticated as mainframes, without the bloat that characterized mainframe system software. Advanced versions of the PDP-11 had an operating system called RT-11 (offered in 1974), which was similar to OS/8, and which further refined the concept of managing data on disks. These were the roots of personal computer operating systems. DEC's role in creating this software ranks with its invention of the minicomputer as major contributions to the creation of personal computing.

Gary Kildall developed PL/M for the Intel 8080. He used an IBM System/360, and PL/M was similar to IBM's PL/I. While working on that project Kildall wrote a small control program for the mainframe's disk drive. "It turned out that the operating system, which was called CP/M for Control Program for Micros, was useful, too, fortunately." Kildall said that PL/M was "the base for CP/M," even though the commands were clearly derived from Digital's, not IBM's software." For example, specifying the drive in use by a letter; giving file names a period and three-character extension; and using the DIR (Directory) command, PIP, and DDT were DEC features carried over without change. CP/M was announced to hobbyists as "similar to DECSYSTEM 10" in an article by Jim Warren in *Dr. Dobb's Journal of Computer Calisthenics and Orthodontia* [sic] in April 1976. Warren was excited by CP/M, stating that it was "well designed, based on an easy-to-use operating system that has been around for a DECade.[sic]" Suggested prices were well under $100, with a complete floppy system that included a drive and a controller for around $800—not cheap, but clearly superior to the alternatives of cassette, paper tape, or any other form of tape. CP/M was the final piece of the puzzle that, when made available, made personal computers a practical reality.

Gary Kildall and his wife, Dorothy McEwen, eased themselves into the commercial software business while he also worked as an instructor at the Naval Postgraduate School in Monterey, California. As interest in CP/M picked up, he found himself writing variations of it for other customers. The publicity in *Dr. Dobb's Journal* led to enough sales to convince him of the potential market for CP/M. In 1976 he quit his job and with Dorothy founded a company, Digital Research (initially Intergalactic Digital Research), whose main product was CP/M.

The next year, 1977, he designed a version with an important difference. IMSAI, the company that had built a "clone" of the Altair, wanted a license to use CP/M for its products. Working with IMSAI employee Glen Ewing, Kildall rewrote CP/M so that only a small portion of it needed to be customized for the specifics of the IMSAI. The rest would be common code that would not have to be rewritten each time a new computer or disk drive came along. He called the specialized code the BIOS—Basic Input/Output System. This change standardized the system software in the same way that the 100-pin Altair bus had standardized hardware. IMSAI's computer system became a standard, with its rugged power supply, room for expansion with plenty of internal slots, external floppy drive, and CP/M.

■ End of the Pioneering Phase, 1977

By 1977 the pieces were all in place. The Altair's design shortcomings were corrected, if not by MITS then by other companies. Microsoft BASIC allowed programmers to write interesting and, for the first time, serious software for these machines. The ethic of charging money for this software gave an incentive to such programmers, although software piracy also became established. Computers were also being offered with BASIC supplied on a read-only-memory (ROM), the manufacturer paying Microsoft a simple royalty fee. (With the start-up codes also in ROM, there was no longer a need for the front panel, with its array of lights and switches.) Eight-inch floppy disk drives, controlled by CP/M, provided a way to develop and exchange software that was independent of particular models. Machines came with standardized serial and parallel ports, and connections for printers, keyboards, and video monitors. Finally, by 1977 there was a strong and healthy industry of publications, software companies, and support groups to bring the novice on board. The personal computer had arrived.

"Code Is Law"

Reading 13

Lawrence Lessig

Like the other authors in this section, Lawrence Lessig believes that technology and society are intertwined. Instead of only examining the past, however, Lessig is especially concerned about the present and future. In tackling a relatively new and still evolving technology—the Internet—Lessig goes as far as to say that computer code will not simply affect our lives, but that it will be as powerful as law in enabling and limiting our actions. He argues that an individual's behavior is influenced by four forms of regulation—law, norms, markets, and what he calls architecture. Architecture is the term he uses for computer code though what he claims applies equally well and more broadly to the built environment. He urges us to act while the technology is still developing so that we can direct it to create a world we want rather than one dictated by those involved in writing code. Lessig's analysis illustrates just one more way in which the social and technical are intertwined and reveals an important implication: In order to get the kind of future we want, we can't focus only on laws, norms and markets. We also must pay attention to technology, technological design, and technical details. Some of Lessig's arguments seem deterministic, but Lessig recognizes that humans have power to control the technology that will ultimately shape us.

▨ Code Is Law

A decade ago, in the spring of 1989, communism in Europe died—collapsed, as a tent would fall if its main post were removed. No war or revolution brought communism to its end. Exhaustion did. Born in its place across Central and Eastern Europe was a new political regime, the beginnings of a new political society.

For constitutionalists (as I am), this was a heady time. I had just graduated from law school in 1989, and in 1991 I began teaching at the University of Chicago. Chicago had a center devoted to the study of the emerging democracies in Central and Eastern Europe. I was a part of that center. Over the next five years I spent more hours on air-planes, and more mornings drinking bad coffee, than I care to remember.

Eastern and Central Europe were filled with Americans telling former Communists how they should govern. The advice was endless and silly. Some of these visitors literally sold constitutions to the emerging constitutional republics; the balance had innumerable half-baked ideas about how the new nations should be governed. These Americans came from a nation where constitutionalism had worked, yet apparently had no clue why.

The center's mission, however, was not to advise. We knew too little to guide. Our aim was to watch and gather data about the transitions and how they progressed. We wanted to understand the change, not direct it.

What we saw was striking, if understandable. Those first moments after communism's collapse were filled with antigovernmental passion—with a surge of anger directed against the state and against state regulation. Leave us alone, the people seemed to say. Let the market and nongovernmental organizations—a new society— take government's place. After generations of communism, this reaction was completely understandable. What compromise could there be with the instrument of your repression?

A certain American rhetoric supported much in this reaction. A rhetoric of libertarianism. Just let the market reign and keep the government out of the way, and freedom and prosperity would inevitably grow. Things would take care of themselves. There was no need, and could be no place, for extensive regulation by the state.

But things didn't take care of themselves. Markets didn't flourish. Governments were crippled, and crippled governments are no elixir of freedom. Power didn't disappear—it simply shifted from the state to mafiosi, themselves often created by the state. The need for traditional state functions—police, courts, schools, health care—didn't magically go away. Private interests didn't emerge to fill the need. Instead, needs were unmet. Security evaporated. A modern if plodding anarchy replaced the bland communism of the previous three generations: neon lights flashed advertisements for Nike; pensioners were swindled out of their life savings by fraudulent stock deals; bankers were murdered in broad daylight on Moscow streets. One system of control had been replaced by another, but neither system was what Western libertarians would call freedom.

At just about the time when this post-communist euphoria was waning—in the mid-1990s—there emerged in the West another "new society," to many just as exciting as the new societies promised in post-communist Europe. This was cyberspace. First in universities and centers of research, and then within society generally, cyberspace became the new target of libertarian utopianism. Here freedom from the state would reign. If not in Moscow or Tblisi, then here in cyberspace would we find the ideal libertarian society.

The catalyst for this change was likewise unplanned. Born in a research project in the Defense Department, cyberspace too arose from the displacement of a certain architecture of control. The tolled, single-purpose network of telephones was displaced by the untolled and multipurpose network of packet-switched data. And thus the old one-to-many architectures of publishing (television, radio, newspapers, books) were supplemented by a world where everyone could be a publisher. People could communicate and associate in ways that they had never done before. The space promised a kind of society that real space could never allow—freedom without anarchy, control without government, consensus without power. In the words of a manifesto that will define our generation: "We reject: kings, presidents and voting. We believe in: rough consensus and running code."[1]

As in post-Communist Europe, first thoughts about cyberspace tied freedom to the disappearance of the state. But here the bond was even stronger than in post-Communist Europe. The claim now was that government *could not* regulate cyberspace, that cyberspace was essentially, and unavoidably, free. Governments could threaten, but behavior could not be controlled; laws could be passed, but they would be meaningless. There was no choice about which government to install—none could reign. Cyberspace would be a society of a very different sort. There would be definition and direction, but built from the bottom up, and never through the direction of a state. The society of this space would be a fully self-ordering entity, cleansed of governors and free from political hacks.

I taught in Central Europe during the summers of the early 1990s; I witnessed the transformation in attitudes about communism that I described at the start of this chapter. And so I felt a bit of déjà vu when in the spring of 1995, I began to teach the law of cyberspace, and saw in my students these very same post-communist thoughts about freedom and government. Even at Yale—not known for libertarian passions—the students seemed drunk with what James Boyle would later call the "libertarian gotcha":[2] no government could survive without the Internet's riches, yet no government could control what went on there. Real-space governments would become as pathetic as the last Communist regimes. It was the withering of the state that Marx had promised, jolted out of existence by trillions of gigabytes flashing across the ether of cyberspace. Cyberspace, the story went, could *only* be free. Freedom was its nature.

But why was never made clear. That *cyberspace* was a place that governments could not control was an idea that I never quite got. The word itself speaks not of freedom but of control. Its etymology reaches beyond a novel by William Gibson (*Neuromancer,* published in 1984) to the world of "cybernetics," the study of control at a distance.[3] Cybernetics had a vision of perfect regulation. Its very motivation was finding a better way to direct. Thus, it was doubly odd to see this celebration of non-control over architectures born from the very ideal of control.

As I said, I am a constitutionalist. I teach and write about constitutional law. I believe that these first thoughts about government and cyberspace are just as misguided as the first thoughts about government after communism. Liberty in cyberspace will not come from the absence of the state. Liberty there, as anywhere, will come from a state of a certain kind.[4] We build a world where freedom can flourish not by removing from society any self-conscious control; we build a world where freedom can flourish by setting it in a place where a particular kind of self-conscious control survives. We build liberty, that is, as our founders did, by setting society upon a certain *constitution.*

But by "constitution" I don't mean a legal text. Unlike my countrymen in Eastern Europe, I am not trying to sell a document that our framers wrote in 1787. Rather, as the British understand when they speak of their constitution, I mean an *architecture*—not just a legal text but a way of life—that structures and constrains social and legal power, to the end of protecting fundamental *values*—principles and ideals that reach beyond the compromises of ordinary politics.

Constitutions in this sense are built, they are not found. Foundations get laid, they don't magically appear. Just as the founders of our nation learned from the anarchy that followed the revolution (remember: our first constitution, the Articles of Confederation, was a miserable failure of do-nothingness), so too are we beginning to see in cyberspace that this building, or laying, is not the work of an invisible hand. There is no reason to believe that the grounding for liberty in cyberspace will simply emerge. In fact, as I will argue, quite the opposite is the case. As our framers learned, and as the Russians saw, we have every reason to believe that cyberspace, left to itself, will not fulfill the promise of freedom. Left to itself, cyberspace will become a perfect tool of control.[5]

Control. Not necessarily control by government, and not necessarily control to some evil, fascist end. But the argument of this book is that the invisible hand of cyberspace is building an architecture that is quite the opposite of what it was at cyberspace's birth. The invisible hand, through commerce, is constructing an architecture that perfects control—an architecture that makes possible highly efficient regulation. As Vernor Vinge warned in 1996, a distributed architecture of regulatory control; as Tom Maddox added, an axis between commerce and the state.[6]

This book is about that change, and about how we might prevent it. When we see the path that cyberspace is on . . . we see that much of the "liberty" present at cyberspace's founding will vanish in its future. Values that we now consider fundamental will not necessarily remain. Freedoms that were foundational will slowly disappear.

If the original cyberspace is to survive, and if values that we knew in that world are to remain, we must understand how this change happens, and what we can do in response. . . . Cyberspace presents something new for those who think about regulation and freedom. It demands a new understanding of how regulation works and of what regulates life there. It compels us to look beyond the traditional lawyer's scope—beyond laws, regulations, and norms. It requires an account of a newly salient regulator.

That regulator is the obscurity in the book's title—*Code*. In real space we recognize how laws regulate—through constitutions, statutes, and other legal codes. In cyberspace we must understand how code regulates—how the software and hardware that make cyberspace what it is *regulate* cyberspace as it is. As William Mitchell puts it, this code is cyberspace's "law."[7] *Code is law.*

This code presents the greatest threat to liberal or libertarian ideals, as well as their greatest promise. We can build, or architect, or code cyberspace to protect values that we believe are fundamental, or we can build, or architect, or code cyberspace to allow those values to disappear. There is no middle ground. There is no choice that does not include some kind of *building*. Code is never found; it is only ever made, and only ever made by us. As Mark Stefik puts it, "Different versions of [cyberspace] support different kinds of dreams. We choose, wisely or not."[8]

My argument is not for some top-down form of control; my claim is not that regulators must occupy Microsoft. A constitution envisions an environment; as Justice Holmes said, it "call[s] into life a being the development of which [can not be] foreseen."[9] Thus, to speak of a constitution is not to describe a one-hundred-day plan. It is instead to identify the values that a space should guarantee. It is not to describe a "government"; it is not even to select (as if a single choice must be made) between bottom-up or top-down control. In speaking of a constitution in cyberspace we are simply asking: What values are protected there? What values will we build into the space to encourage certain forms of life?

The "values" here are of two sorts—substantive and structural. In the American tradition, we worried about the second first. The framers of the Constitution of 1787 (enacted without a Bill of Rights) were focused on structures of government. Their aim was to ensure that a particular government (the federal government) did not become too powerful. And so they built into its design checks on the power of the federal government and limits on its reach over the states.

Opponents of that Constitution insisted that more checks were needed, that the Constitution needed to impose substantive limits on government's power as well as structural limits. And thus the Bill of Rights was born. Ratified in 1791, the Bill of Rights promised that the federal government will not remove certain protections—of speech, privacy, and due process. And it guaranteed that the commitment to these substantive values will remain despite the passing fancy of normal government. These values were to be entrenched, or embedded, in our constitutional design; they can be changed, but only by changing the Constitution's design.

These two kinds of protection go together in our constitutional tradition. One would have been meaningless without the other. An unchecked structure could easily have overturned the substantive protections expressed in the Bill of Rights, and without substantive protections, even a balanced and reflective government could have violated values that our framers thought fundamental.

We face the same questions in constituting cyberspace, but we have approached them from an opposite direction. Already we are struggling with substance: Will cyberspace promise privacy or access? Will it preserve a space for free speech? Will it facilitate free and open trade? These are choices of substantive value.

But structure matters as well. What checks on arbitrary regulatory power can we build into the design of the space? What "checks and balances" are possible? How do we separate powers? How do we ensure that one regulator, or one government, doesn't become too powerful? . . .

Theorists of cyberspace have been talking about these questions since its birth.[10] But as a culture, we are just beginning to get it. We are just beginning to see why the architecture of the space matters—in particular, why the *ownership* of that architecture matters. If the code of cyberspace is owned, . . . it can be controlled; if it is not owned, control is much more difficult. The lack of ownership, the absence of properly, the inability to direct how ideas will be used—in a word, the presence of a commons—is key to limiting, or checking, certain forms of governmental control.

One part of this question of ownership is at the core of the current debate between open and closed source software. In a way that the American founders would have instinctively understood, "free software" or "open source software"—or "open code," to (cowardly) avoid taking sides in a debate I describe later—is itself a check on arbitrary power. A structural guarantee of constitutionalized liberty, it functions as a type of separation of powers in the American constitutional tradition. It stands alongside substantive protections, like freedom of speech or of the press, but its stand is more fundamental. . . . The first intuition of our r founders was right: structure builds substance. Guarantee the structural (a space in cyberspace for open code), and (much of) the substance will take care of itself.

* * *

Given our present tradition in constitutional law and our present faith in representative government, are we able to respond collectively to the changes I will have described?

My strong sense is that we are not. We are at a stage in our history when we urgently need to make fundamental choices about values, but we trust no institution of government to make such choices. Courts cannot do it, because as a legal culture we don't want courts choosing among contested matters of values, and Congress should not do it because, as a political culture, we so deeply question the products of ordinary government.

Change is possible. I don't doubt that revolutions remain in our future; the open code movement is just such a revolution. But I fear that it is too easy for the government to dislodge these revolutions, and that too much will be at stake for it to allow the revolutionaries to succeed. Our government has already criminalized the core ethic of this movement, transforming the meaning of *hacker* into something quite alien to its original sense. This, I argue, is only the start.

Things could be different. They are different elsewhere. But I don't see how they could be different for us just now. This no doubt is a simple confession of the limits of my own imagination. I would be grateful to be proven wrong. I would be grateful to watch as we relearn—as the citizens of the former Communist republics are learning—how to escape our disabling ideas about the possibilities for governance.

* * *

■ What Things Regulate

John Stuart Mill was an Englishman, though one of the most influential political philosophers in America in the nineteenth century. His writings ranged from important work on logic to a still striking text, *The Subjection of Women*. But his continuing influence comes from a relatively short book titled *On Liberty*. Published in 1859, this powerful argument for individual liberty and diversity of thought represents an important view of liberal and libertarian thinking in the second half of the nineteenth century.

"Libertarian," however, has a specific meaning for us. It associates with arguments against government.[11] Government, in the modern libertarian's view, is the threat to liberty; private action is not. Thus, the good libertarian is focused on reducing government's power. Curb the excesses of government, the libertarian says, and you will have ensured freedom for your society.

Mill's view was not so narrow. He was a defender of liberty and an opponent of forces that suppressed it. But those forces were not confined to government. Liberty, in Mill's view, was threatened as much by norms as by government, as much by stigma and intolerance as by the threat of state punishment. His objective was to argue against these private forces of coercion. His work was a defense against liberty-suppressing norms, because in England at the time these were the real threat to liberty.

Mill's method is important, and it should be our own. It asks, What is the threat to liberty, and how can we resist it? It is not limited to asking. What is the threat to liberty *from government?* It understands that more than government can threaten liberty, and that sometimes this something more can be private rather than state action. Mill was not so concerned with the source. His concern was with liberty.

Threats to liberty change. In England norms may have been the problem in the late nineteenth century; in the United States in the first two decades of the twentieth century it was state suppression of speech.[12] The labor movement was founded on the idea that the market is sometimes a threat to liberty—not just because of low wages, but also because the market form of organization itself disables a certain kind of freedom.[13] In other societies, at other times, the market is the key, not the enemy, to liberty.

Thus, rather than think of an enemy in the abstract, we should understand the particular threat to liberty that exists in a particular time and place. And this is especially true when we think about liberty in cyberspace. For my argument is that cyberspace teaches a new threat to liberty. Not new in the sense that no theorist has conceived of it before. Others have.[14] But new in the sense of newly urgent. We are coming to understand a newly powerful regulator in cyberspace, and we don't yet understand how best to control it.

This regulator is code—or more generally, the "built environment" of social life, its architecture.[15] And if in the middle of the nineteenth century it was norms that threatened liberty, and at the start of the twentieth state power that threatened liberty, and during much of the middle twentieth century the market that threatened liberty, my argument is that we understand how in the late twentieth century, and into the twenty-first, it is a different regulator—code—that should be our concern.

But it is not my aim to say that this should be our new single focus. My argument is not that there is a new single enemy different from the old. Instead, I believe we need a more general understanding of how regulation works. One that focuses on more than the single influence of any one force such as government, norms, or the market, and instead integrates these factors into a single account.

This chapter is a step toward that more general understanding.[16] It is an invitation to think beyond the narrow threat of government. The threats to liberty have never come solely from government, and the threats to liberty in cyberspace certainly will not.

■ A Dot's Life

There are many ways to think about constitutional law and the limits it may impose on government regulation. I want to think about it from the perspective of someone who is regulated or constrained. That someone regulated is represented by this (pathetic) dot—a creature (you or me) subject to the different constraints that might regulate it. By describing the various constraints that might bear on this individual, I hope to show you something about how these constraints function together.

Here then is the dot.

How is this dot "regulated"?

Let's start with something easy: smoking. If you want to smoke, what constraints do you face? What factors *regulate* your decision to smoke or not?

One constraint is legal. In some places at least, laws regulate smoking—if you are under eighteen, the law says that cigarettes cannot be sold to you. If you are under twenty-six, cigarettes cannot be sold to you unless the seller checks your ID. Laws also regulate where smoking is permitted—not in O'Hare Airport, on an airplane, or in an elevator, for instance. In these two ways at least, laws aim to direct smoking behavior. They operate as a kind of constraint on an individual who wants to smoke.[17]

But laws are not the most significant constraints on smoking. Smokers in the United States certainly feel their freedom regulated, even if only rarely by the law. There are no smoking police, and smoking courts are still quite rare. Rather, smokers in America are regulated by norms. Norms say that one doesn't light a cigarette in a private car without first asking permission of the other passengers. They also say, however, that one needn't ask permission to smoke at a picnic. Norms say that others can ask you to stop smoking at a restaurant, or that you never smoke during a meal.

European norms are savagely different. There the presumption is in the smoker's favor; vis-à-vis the smoker, the norms are laissez-faire. But in the States the norms effect a certain constraint, and this constraint, we can say, *regulates* smoking behavior.

Law and norms are still not the only forces regulating smoking behavior. The market too is a constraint. The price of cigarettes is a constraint on your ability to smoke. Change the price, and you change this constraint. Likewise with quality. If the market supplies a variety of cigarettes of widely varying quality and price, your ability to select the kind of cigarette you want increases; increasing choice here reduces constraint.

Finally, there are the constraints created, we might say, by the technology of cigarettes, or by the technologies affecting their supply.[18] Unfiltered cigarettes present a greater constraint on smoking than filtered cigarettes if you are worried about your health. Nicotine-treated cigarettes are addictive and therefore create a greater constraint on smoking than untreated cigarettes. Smokeless cigarettes present less of a constraint because they can be smoked in more places. Cigarettes with a strong odor present more of a constraint because they can be smoked in fewer places. In all of these ways, how the cigarette *is* affects the constraints faced by a smoker. How it is, how it is designed, how it is built—in a word, its *architecture.*

Thus, four constraints regulate this pathetic dot—the law, social norms, the market, and architecture—and the "regulation" of this dot is the sum of these four constraints. Changes in any one will affect the regulation of the whole. Some constraints will support others; some may undermine others. A complete view, however, should consider them together.

So think of the four together like this:

In this drawing, each oval represents one kind of constraint operating on our pathetic dot in the center. Each constraint imposes a different kind of cost on the dot for engaging in the relevant behavior—in this case, smoking. The cost from norms is different from the market cost, which is different from the cost from law and the cost from the (cancerous) architecture of cigarettes.

The constraints are distinct, yet they are plainly interdependent. Each can support or oppose the others. Technologies can undermine norms and laws; they can also support them. Some constraints make others possible; others make some impossible. Constraints work together, though they function differently and the effect of each is distinct. Norms constrain through the stigma that a community imposes; markets constrain through the price that they exact; architectures constrain through the physical burdens they impose; and law constrains through the punishment it threatens.

We can call each constraint a "regulator," and we can think of each as a distinct modality of regulation. Each modality has a complex nature, and the interaction among these four is hard to describe. I've worked through this complexity more completely in the appendix. But for now, it is enough to see that they are linked and that, in a sense, they combine to produce the regulation to which our pathetic dot is subject in any given area.

The same model describes the regulation of behavior in cyberspace.

Law regulates behavior in cyberspace. Copyright law, defamation law, and obscenity laws all continue to threaten *ex post* sanction for the violation of legal rights. How well law regulates, or how efficiently, is a different question: in some cases it does so more efficiently, in some cases less. But whether better or not, law continues to threaten a certain consequence if it is defied. Legislatures enact;[19] prosecutors threaten;[20] courts convict.[21]

Norms also regulate behavior in cyberspace. Talk about democratic politics in the alt.knitting newsgroup, and you open yourself to flaming; "spoof" someone's identity in a MUD, and you may find yourself "toaded";[22] talk too much in a discussion list, and you are likely to be placed on a common bozo filter. In each case, a set of understandings constrain behavior, again through the threat of *ex post* sanctions imposed by a community.

Markets regulate behavior in cyberspace. Pricing structures constrain access, and if they do not, busy signals do. (AOL learned this quite dramatically when it shifted from an hourly to a flat rate pricing plan.)[23] Areas of the Web are beginning to charge for access, as online services have for some time. Advertisers reward popular sites; online services drop low-population forums. These behaviors are all a function of market constraints and market opportunity. They are all, in this sense, regulations of the market.

And finally, an analog for architecture regulates behavior in cyberspace—*code*. The software and hardware that make cyberspace what it is constitute a set of constraints on how you can behave. The substance of these constraints may vary, but they are experienced as conditions on your access to cyberspace. In some places (online services such as AOL, for instance) you must enter a password before you gain access; in other places you can enter whether identified or not.[24] In some places the transactions you engage in produce traces that link the transactions (the "mouse droppings") back to you; in other places this link is achieved only if you want it to be.[25] In some places you can choose to speak a language that only the recipient can hear (through encryption);[26] in other places encryption is not an option.[27] The code or software or architecture or protocols set these features; they are features selected by code writers; they constrain some behavior by making other behavior possible, or impossible. The code embeds certain values or makes certain values impossible. In this sense, it too is regulation, just as the architectures of real-space codes are regulations.

As in real space, then, these four modalities regulate cyberspace. The same balance exists. As William Mitchell puts it (though he omits the constraint of the market); Architecture, laws, and customs maintain and represent whatever balance has been struck [in real space]. As we construct and inhabit cyberspace communities, we will have to make and maintain similar bargains—though they will be embodied in software structures and electronic access controls rather than in architectural arrangements.[28]

Laws, norms, the market, and architectures interact to build the environment that "Netizens" know. The code writer, as Ethan Katsh puts it, is the "architect."[29]

But how can we "make and maintain" this balance between modalities? What tools do we have to achieve a different construction? How might the mix of real-space values be carried over to the world of cyberspace? How might the mix be changed if change is desired?

Notes

1. Paulina Borsook, "How Anarchy Works," *Wired* 110 (October 1995): 3.10, available at http://www.wired.com/wired/archive/3.10/ietf_pr.html (visited May 30, 1999), quoting Net-lander, David Clark.
2. James Boyle, talk at Telecommunications Policy Research Conference (TPRC), Washington, D.C., September 28, 1997. David Shenk discusses the libertarianism that cyberspace inspires (as well as other, more fundamental problems with the age) in a brilliant cultural how-to book that responsibly covers both the technology and the libertarianism; see *Data Smog: Surviving the Information Glut* (San Francisco, Harper Edge, 1997), esp. 174–77. The book also describes technorealism, a responsive movement that advances a more balanced picture of the relationship between technology and freedom.
3. See Kevin Kelley, *Out of Control: The New Biology of Machines, Social Systems, and the Economic World* (Reading, Mass.: Addison-Wesley, 1994), 119.
4. As Stephen Holmes has put it, "Rights depend upon the competent exercise of . . . legitimate public power. . . . The largest and most reliable human rights organization is the liberal state. . . . Unless society is politically well organized, there will be no individual liberties and no civil society"; "What Russia Teaches Us Now: How Weak States Threaten Freedom," *American Prospect* 33 (1997): 30, 33.
5. This is a dark picture, I confess, and it contrasts with the picture of control drawn by Andrew Shapiro in *The Control Revolution* (New York: Public Affairs, 1999). As I discuss later, however, the difference between Shapiro's view and my own turns on the extent to which architectures enable top-down regulation. In my view, one highly probable architecture would enable greater regulation than Shapiro believes is likely.
6. See "We Know Where You Will Live," Computers, Freedom, and Privacy Conference, March 30, 1996, audio link available at http://www-swiss.ai.mit.edu/projects/mac/cfp96/plenary-sf.html.
7. See William J. Mitchell, *City of Bits: Space, Place, and the Infobahn* (Cambridge, Mass.: MIT Press, 1995), 111. In much of this book, I work out Mitchell's idea, though I drew the metaphor from others as well. Ethan Katsh discusses this notion of software worlds in "Software Worlds and the First Amendment: Virtual Doorkeepers in Cyberspace," *University of Chicago Legal Forum* (1996): 335, 338. Joel Reidenberg discusses the related notion of "lex informatica" in "Lex Informatica: The Formulation of Information

Policy Rules Through Technology," *Texas Law Review* 76 (1998): 553. I have been especially influenced by James Boyle's work in the area. I discuss his book in chapter 9, but see also "Foucault in Cyberspace: Surveillance, Sovereignty, and Hardwired Censors," *University of Cincinnati Law Review* 66 (1997): 177. For a recent and powerful use of the idea, see Shapiro, *The Control Revolution.* Mitch Kapor is the father of the mane "architecture is politics" within cyberspace talk. I am indebted to him for this.

8. Mark Stefik, "Epilogue: Choices and Dreams," in *Internet Dreams: Archetypes, Myths, and Metaphors,* edited by Mark Stefik (Cambridge, Mass.: MIT Press, 1996), 390.

9. *Missouri v Holland,* 252 US 416, 433 (1920).

10. Richard Stallman, for example, organized resistance to the emergence of passwords at MIT. Passwords are an architecture that facilitates control by excluding users not "officially sanctioned." Steven Levy, *Hackers* (Garden City, N.V.: Anchor Press/Doubleday, 1984), 416–17.

11. Or more precisely, against a certain form of government regulation. The more powerful libertarian arguments against regulation in cyberspace are advanced, for example, by Peter Huber in *Law and Disorder in Cyberspace: Abolish the FCC and Let Common Law Rule the Telecosm.* Huber argues against agency regulation and in favor of regulation by the common law. See also Thomas Hazlett in "The Rationality of U.S. Regulation of the Broadcast Spectrum," *Journal of Law and Economics* 33 (1990): 133, 133–39. For a lawyer, it is hard to understand precisely what is meant by "the common law." The rules of the common law are many, and the substantive content has changed. There is a common law process, which lawyers like to mythologize, in which judges make policy decisions in small spaces against the background of binding precedent. It might be this that Huber has in mind, and if so, there are, of course, benefits to this system. But as he plainly understands, it is a form of *regulation* even if it is constituted differently.

12. The primary examples are the convictions under the 1917 Espionage Act; see, for example, *Schenck v United States,* 249 US 47 (1919) (upholding conviction for distributing a leaflet attacking World War I conscription); *Frohwerk v United States,* 249 US 204 (1919) (upholding conviction based on newspaper alleged to cause disloyalty); *Debs v United States,* 249 US 211 (1919) (conviction upheld for political speech said to cause insubordination and disloyalty).

13. See, for example, the work of John R. Commons, *Legal Foundations of Capitalism* (1924), 296–98, discussed in Herbert Hovenkamp, *Enterprise and American Law, 1836–1937* (Cambridge, Mass.: Harvard University Press, 1991), 235; see also John R. Commons, *Institutional Economics: Its Place in Political Economy* (1934).

14. The general idea is that the tiny corrections of space enforce a discipline, and that this discipline is an important regulation. Such theorizing is a tiny part of the work of Michel Foucault; see *Discipline and Punish: The Birth of the Prison* (1979), 170–77, though his work generally inspires this perspective. It is what Oscar Gaudy speaks about in *The Panoptic Sort: A Political Economy of Personal Information* (Boulder, Colo.: Westview Press, 1993), 23. David Brin makes the more general point that I am arguing—that the threat to liberty is broader than a threat by the state; see David Brin, *The Transparent Society: Will Technology Force Us to Choose between Privacy and Freedom?* (New York: Basic Books, 1990), 110.

15. See, for example, *The Built Environment: A Creative Inquiry into Design and Planning,* edited by Tom J. Bartuska and Gerald L. Young (Menlo Park, Calif.: Crisp Publications, 1994); *Preserving the Built Heritage: Tools for Implementation,* edited by J. Mark Schuster et al. (Hanover, N.H.: University Press of New England, 1997). In design theory, the notion I am describing accords with the tradition of Andres Duany and Elizabeth Plater-Zyberk; see, for example, William Lennertz. "Town-Making Fundamentals," in *Towns and Town-Making Principles,* edited by Andres Duany and Elizabeth Plater-Zyberk (New York: Rizzoli, 1991): "The work of . . . Duany and . . . Plater-Zyberk begins with the recognition that design affects behavior. [They] see the structure and function of a community as interdependent. Because of this, they believe a designer's decisions will permeate the lives of residents not just visually but in the way residents live. They believe design structures functional relationships, quantitatively and qualitatively, and that it is a sophisticated tool whose power exceeds its cosmetic attributes" (21).

16. Elsewhere I've called this the "New Chicago School"; see Lawrence Lessig, "The New Chicago School," *Journal of Legal Studies* 27 (1998): 661. It is within the "tools approach" to government action

(see John de Monchaux and J. Mark Schuster, "Five Things to Do," in Schuster, *Preserving the Built Heritage,* 3), but it describes four tools whereas Schuster describes five. I develop the understanding of the approach in the appendix to this book.

17. See generally *Smoking Policy: Law, Politics, and Culture,* edited by Robert L. Rabin and Stephen D. Sugarman (New York: Oxford University Press, 1993); Lawrence Lessig, "The Regulation of Social Meaning." *University of Chicago Law Review* 62 (1995): 943, 1025–34; Cass R. Sunstein, "Social Norms and Social Roles," *Columbia Law Review* 96 (1996): 903.

18. These technologies are themselves affected, no doubt, by the market. Obviously, these constraints could not exist independently of each other but affect each other in significant ways.

19. The ACLU lists twelve states that passed Internet regulations between 1995 and 1997; see http://www .aclu.org/issues/cyber/censor/stbills.html#bills (visited May 31, 1999).

20. See, for example, the policy of the Minnesota attorney general on the jurisdiction of Minnesota over people transmitting gambling information into the state; available at http://www.ag.state.mn.us/home/ consumer/consumernews/OnlineScams/memo.html (visited May 31, 1999).

21. See, for example, *Playboy Enterprises v Chuckleberry Publishing, Inc.,* 939 FSupp 1032 (SDNY 1996); *United States v Thomas,* 74 F3d 1153 (6th Cir 1996); *United States v Miller,* 166 F3d 1153 (11th Cir 1999); *United States v Large,* 166 F3d 516 (2d Cir 1999); *United States v Whiting,* 165 F3d 631 (8th Cir 1999); *United States v Hibbler,* 159 F3d 233 (6th Cir 1998); *United States v Fellows,* 157 F3d 1197 (9th Cir 1998); *United States v Simpson,* 152 F3d 1241 (10th Cir 1998); *United States v Hall,* 142 F3d 988 (7th Cir 1998); *United States v Hackings,* 129 F3d 1069 (9th Cir 1997); *United States v Lacy,* 119 F3d 742 (9th Cir 1997); *United States v Smith,* 47 MJ 588 (CrimApp 1997); *United States v Ownby,* 926 FSupp 558 (WDVa 1996).

22. See Julian Dibbell, "A Rape in Cyberspace," *Village Voice,* December 23, 1993, 36.

23. See, for example, "AOL Still Suffering but Stock Price Rises," *Network Briefing,* January 31, 1997; David S. Hilzenrath, "'Free' Enterprise, Online Style; AOL, CompuServe, and Prodigy Settle FTC Complaints," *Washington Post,* May 2, 1997, G1; "America Online Plans Better Information About Price Changes," *Wall Street Journal,* May 29, 1998, B2; see also Swisher, *Aol.com,* 206–8.

24. USENET postings can be anonymous; see Henry Spencer and David Lawrence, *Managing USENET* (Sebastopol, Calif.: O'Reilly & Associates, 1998), 366–67.

25. Web browsers make this information available, both in real time and archived in a cookie file; see http:// www.cookiecentral.com/faq.htm (visited May 31, 1999). They also permit users to turn this tracking feature off.

26. PGP is a program to encrypt messages that is offered both commercially and free.

27. Encryption, for example, is illegal in some international contexts; see Stewart A. Baker and Paul R. Hurst, *The Limits of Trust: Cryptography, Governments, and Electronic Commerce* (Boston: Kluwer Law International, 1998) 130–6.

28. William J. Mitchell, *City of Bits: Space, Place, and the Infobahn* (Cambridge, MA: MIT Press, 1995), 159.

29. See M. Ethan Katsh, "Software Worlds and the First Amendment: Virtual Doorkeepers in Cyberspace," *University of Chicago legal Forum* (1996): 335, 340. "If a comparison to the physical world is necessary, one might say that the software designer is the architect, the builder, and the contractor, as well as the interior decorator."

Section 4

Genetic Engineering and the Environment

Rosalind Franklin and the Double Helix

Reading 14

Lynne Osman Elkin

Although she made essential contributions toward elucidating the structure of DNA, Rosalind Franklin is known to many only as seen through the distorting lens of James Watson's book, *The Double Helix*.

In 1962, James Watson, then at Harvard University, and Cambridge University's Francis Crick stood next to Maurice Wilkins from King's College, London, to receive the Nobel Prize in Physiology or Medicine for their "discoveries concerning the molecular structure of nucleic acids and its significance for information transfer in living material." Watson and Crick could not have proposed their celebrated structure for DNA as early in 1953 as they did without access to experimental results obtained by King's College scientist Rosalind Franklin. Franklin had died of cancer in 1958 at age 37, and so was ineligible to share the honor. Her conspicuous absence from the awards ceremony—the dramatic culmination of the struggle to determine the structure of DNA—probably contributed to the neglect, for several decades, of Franklin's role in the DNA story. She most likely never knew how significantly her data influenced Watson and Crick's proposal.

■ Capsule Biography

Franklin, was born 25 July 1920 to Muriel Waley Franklin and merchant banker Ellis Franklin, both members of educated and socially conscious Jewish families. They were a close immediate family, prone to lively discussion and vigorous debates at which the politically liberal, logical, and determined Rosalind excelled: She would even argue with her assertive, conservative father. Early in life, Rosalind manifested the creativity and drive characteristic of the Franklin women, and some of the Waley women, who were expected to focus their education, talents, and skills on political, educational, and charitable forms of community service. It was thus surprising when young Rosalind expressed an early fascination with physics and chemistry classes at the academically rigorous St. Paul's Girls' School in London, and unusual that she earned a bachelor's degree in natural sciences with a specialty in physical chemistry. The degree was earned at Newnham College, Cambridge in 1941.

From 1942 to 1946, Franklin did war-related graduate work with the British Coal Utilization Research Association. That work earned her a PhD from Cambridge in 1945, and an offer to join the Laboratoire Central des Services Chimiques de l'Etat in Paris. She worked there, from 1947 to 1950, with Jacques Mering and became proficient at applying x-ray diffraction techniques to imperfectly crystalline matter such as coal. In the period 1946–49, she published five landmark coal-related papers, still cited today, on graphitizing and nongraphitizing carbons. By 1957, she had published an additional dozen articles on carbons other than coals. Her papers changed the way physical chemists view the microstructure of coals and related substances.

Franklin made many friends in the Paris laboratory and often hiked with them on weekends. She preferred to live on her own modest salary and frustrated her parents by continually refusing to accept money from them. She excelled at speaking French and at French cooking and soon became more comfortable with intellectual and egalitarian "French ways" than with conventional English middle-class customs. Consequently, she did not fit in well at King's College, where she worked on DNA from 1951 to 1953. Franklin chose to leave King's and, in the spring of 1953, moved to Birkbeck College. Many of the students there were evening students

Lynne Elkin *(lelkin@csuhayward.edu) is a professor of biological sciences at California State University, Hayward. She welcomes responses and inquiries about issues pertaining to the story of Rosalind Franklin and DNA structure.*

who worked during the day, and Franklin was impressed with their dedication. After the move to Birkbeck, she began her celebrated work with J. Desmond Bernal on RNA viruses like tobacco mosaic virus (TMV). She was a cautious scientist who began to trust her intuition more as she matured (see box 1 on page 45). She published 14 papers about viruses between 1955 and 1958, and completed the research for three others that colleague Aaron Klug submitted for publication after her death.

In his obituary for Franklin, Bernal described her as a "recognized authority in industrial physico-chemistry." In conclusion, he wrote, "As a scientist, Miss Franklin was distinguished by extreme clarity and perfection in everything she undertook. Her photographs are among the most beautiful of any substances ever taken."[1]

■ Discovery of Two Forms for DNA

Franklin's most famous and controversial work yielded critical data that Watson and Crick used to determine DNA's structure. DNA is a double-helical molecule roughly in the form of a spiral staircase. The double-helical molecule, consisting of two unbranched polynucleotide chains, is best visualized by imagining it straightened into a ladder. The side rails of the ladder are each made up of alternating sugar and phosphate groups, linked by so-called 3′ or 5′ phosphodiester bonds. The sequence of the atoms in each rail runs in opposite directions, so the two sides of the molecular backbone are often described as antiparallel to each other. The rungs of the ladder consist of specific hydrogen-bonded horizontal pairs of nitrogenous bases that are attached to the deoxyribose sugars in the backbone's side rails. Birkbeck's Sven Furberg, who studied both the nucleoside and nucleotide of one of the bases, cytosine, in 1949, discovered that the base would be perpendicular to the sugar: The result can be extrapolated to hydrated DNA.[2]

The pairs of nitrogenous bases that make up the rungs are in the keto, as opposed to enol, tautomeric configurations. (The two are distinguished by the locations of hydrogen atoms available for hydrogen bonding.) The smaller single-ringed pyrimidines, cytosine (C) and thymine (T), are always paired with larger double-ringed purines, guanine (G) and adenine (A). Indeed, the consistent pairing of G with C and of A with T, as first proposed by Watson, explains the identical size of the ladder rungs and also Erwin Chargaff's 1952 observation that G and C (and likewise A and T) are always present in DNA in approximately equal amounts. The consistent pairings, along with the irregular linear vertical sequence of the bases, underlie DNA's genetic capacity.

In an experiment carried out shortly after she arrived at King's, Franklin identified two distinct configurations, called by her the A and B forms, in which DNA could exist. Her work, first presented in an internal King's seminar in November 1951 and published in *Nature* in 1953, was essential for determining the structure of DNA. Researchers working prior to Franklin's discovery invariably had to deal with confusing x-ray diffraction patterns that resulted from a mixture of the A and B forms.

The drier crystalline A form contains about 20% water by weight and is optimally produced at about 75% relative humidity. Cation (for example, Na^+) bridges between ionized phosphates are probably responsible for intermolecular linking in the crystalline structure. The less ordered, fully hydrated, paracrystalline B form— typically the configuration that occurs in vivo—is obtained from the crystalline A form when DNA fibers absorb water in excess of 40% of their weight. Optimal production of the B form occurs at approximately 90% relative humidity. Extra hydration makes it easier for the molecule to assume the lowest-energy, helical configuration. It also keeps the two helical backbone chains farther apart than in the A form and elongates the molecule by about 30% until the B form appears with its bases oriented perpendicular to the fiber axis.

Franklin slowly and precisely hydrated then dehydrated her DNA sample to obtain her best pictures of the A form. To get her samples, though, she had to extract DNA fibers from a gel-like undenatured DNA sample that Wilkins had acquired from Rudolf Signer of the University of Berne in Switzerland. Franklin pulled exceptionally thin single fibers and controlled the humidity in her specimen chamber by bubbling hydrogen gas through salt solutions and then flooding the chamber with the humid gas that resulted. Franklin's PhD student, Raymond Gosling, told me that the chamber leaked so much hydrogen gas that he was afraid they would blow themselves up accidentally and take half of King's College with them.

Franklin's careful treatment during the transformation from crystalline A-form to hydrated B-form DNA resulted in such a drastic size change that, according to Gosling, the elongating specimen practically "leaped off the stage." After designing a tilting microfocus camera and developing a technique for improving the orientation of her DNA fibers in the camera's collimated beam, Franklin took x-ray diffraction photographs of the B form.

Franklin's B-form data, in conjunction with cylindrical Patterson map calculations that she had applied to her A-form data, allowed her to determine DNA's density, unit-cell size, and water content. With those data, Franklin proposed a double-helix structure with precise measurements for the diameter, the separation between each of the coaxial fibers along the fiber axis direction, and the pitch of the helix.[3]

The resolution of the B-form photograph #51 allowed Franklin to determine that each turn of the helix in the B form is 34 Å long and contains 10 base pairs separated by 3.4 Å each,[3] in accordance with less precise data obtained by William Astbury and Florence Bell in 1938.[4] Wilkins, photographing living sperm cells in 1952, obtained an X-shaped B-form diffraction pattern similar to Franklin's. Her photograph, though, showed much more detail.

■ Additional Contributions

The cylindrical Patterson map calculations that Franklin applied to the A-form of DNA were the first such calculations applied to any molecule. They confirmed her suggestions that the hydrophilic sugar phosphates form the external backbone of the DNA molecule and that the hydrophobic base pairs are protected inside that backbone from the cell's aqueous environment. The calculations also allowed her to deduce that the A-form helix has two antiparallel chains. With Gosling, Franklin provided details of the physical distortion accompanying the dehydration transformation from B-form to A-form DNA.[5] She also showed that the bases of the A form are tilted and curved slightly, and that 11 pairs of bases are compacted within a repeat distance of 28.1 Å.

In her section of the 1952 King's Medical Research Council (MRC) report, Franklin gave quantitative measurements for the interphosphate distances and discussed the external placement of the phosphates. Her presentation was instrumental in getting Watson and Crick to abandon their earlier attempts at placing the bases on the outside of their model. Initially, they (and Linus Pauling, too) mistakenly thought that the bases would have to be externally accessible in order to pass genetic information.

In May 1952, Franklin presented her clearest evidence of the helical backbone, with her diffraction photograph #51. Although she did not yet realize how the nitrogenous bases are paired or that the helical backbone rails of B-form DNA are antiparallel, her notebook entries starting in January 1951 clearly show that she was making significant progress toward solving those two final aspects of DNA structure. After reading an article by June Broomhead,[6] and studying other related papers, she had used the keto configuration for at least three of the four bases. She was aware both of Jerry Donohue's work concerning tautomeric forms of bases and of Chargaff's work.

Astbury and Bell's earlier, less clear diffraction photographs and later data of Wilkins suggested some of the data that Franklin derived from her photograph #51. But Franklin's results were much more precise than the Astbury and Bell data, which showed neither an X pattern nor layer lines. Astbury and Bell themselves described their results as "still rather obscure." After Oxford crystallographer Dorothy Hodgkin helped her to eliminate two of three possibilities she had calculated, Franklin described the correct crystallographic space group for DNA in the 1952 MRC report.

Only after Crick obtained Franklin's data—his thesis adviser, Max Perutz, agreed to give him a copy of the 1952 report and Watson had seen photograph #51—was he sufficiently convinced to start constructing the backbone of the successful DNA model. He recognized the similarity of the space group Franklin had calculated to that of his thesis molecule, hemoglobin, and immediately deduced that there would be an antiparallel orientation between the two DNA coaxial fibers. Within one week, he started modeling the correct backbone in a manner compatible with Franklin's data. On several occasions, Crick has acknowledged that the data and conclusions in the 1952 report were essential.

■ Franklin's 17 March 1953 Draft

On 18 March 1953, Wilkins penned a letter acknowledging receipt of the Watson and Crick manuscript that described the structure of DNA. A day earlier, Franklin, who was preparing to leave for Birkbeck, polished an already written draft manuscript outlining her conclusions about the double-helix backbone chain of B-form DNA.[7] Franklin only slightly modified her draft to prepare her April 1953 *Nature* paper, which appeared as the third in a series that led off with the famous Watson and Crick proposal. Partly as a consequence of its placement, Franklin's paper seemed merely to support Watson and Crick's work. But Franklin's data played far more than just a supporting role—as early as 1968, Watson's *The Double Helix* tells us so.

Ironically, despite its negative portrayal of Franklin, *The Double Helix* was what first brought widespread attention to Franklin's key contributions to the Watson and Crick proposal. The book describes how Watson and Crick built their first, and incorrect, model right after Watson inaccurately reported Franklin's November 1951 seminar data to Crick. It also details how, after 13 months of inactivity, they built their correct model once Wilkins showed Franklin's photograph #51 to Watson, and Perutz showed Crick the 1952 MRC report.

The importance of Franklin's work to the discovery of DNA structure has not been well documented until recently for a variety of reasons too long to discuss here. Relevant issues include women's being underrepresented in historical accounts, although several authors have striven to correct that imbalance;[8] Watson and Crick's routinely citing the more senior Wilkins before Franklin; and Wilkins's repeating much of Franklin's work. In addition, Wilkins, not Franklin, was nominated for membership in the Royal Society even though, at the time of his nomination, Franklin was famous for her TMV accomplishments.

■ Conflict within the King's MRC

Franklin was an outstanding and accomplished scientist—a fascinating individual with a strong personality who made a lasting impression on almost everyone she met. Throughout her career, she routinely ate lunch amicably with both male and female colleagues and most of her acquaintances liked her. Her numerous lifelong friends thought her bright, fascinating, witty, and fun. Most of her lunchtime colleagues at King's would agree with that description, but only as it pertained to her lunchtime persona. When Rosalind headed for the laboratory, she shed almost every vestige of lightheartedness as she focused exclusively on her work.

Furthermore, in what amounted almost to social heresy in England, en route to her laboratory she typically bypassed the morning coffeepot and afternoon tea in favor of a direct assault on her work. Franklin was considered by King's colleagues to be "too French" in her dress, in her intellectual interests, and in her temperament. She was exceedingly direct, intent, and serious, with the tendency to leap into passionate debate. She could be assertive, uniquely stubborn, argumentative, and abrasive to the point that colleagues, especially Wilkins, sometimes found her unpleasant.

On the rare occasions when Franklin departed from her typical behavior, King's colleagues usually did not even notice—they took her seriously at all times. That became especially important when she created a "death of the helix" funeral invitation as a joke after obtaining some data indicating that A-form DNA is nonhelical (see box 2 on page 46).

Whereas Franklin was quick, intense, assertive, and directly confrontational, Wilkins was exceedingly shy, indirect, and slowly calculating to the point of appearing plodding. Almost all testimony from King's staff indicates that any blame for their perpetual conflict needs to be shared. John Randall, director of the MRC, also deserves some share of the blame. Without informing Wilkins, he wrote a letter to Franklin assigning DNA structural studies to her. He also did not warn Franklin about Wilkins's continuing interest in DNA.

Randall and Wilkins did make some important accommodations for Franklin. They hired her into a senior position based on her expertise. They gave her an excellent laboratory, the highest quality DNA, and a decent budget. However, numerous MRC women, although very well treated for that era, did not receive precisely equal treatment with men. Taken as a whole, the King's MRC women did not rank quite as highly as the men. Also, Gosling told me that women were not allowed upstairs for after-lunch coffee in the smoking room, "a wonderful long room," he explained, "with window seats." That exclusion undoubtedly cut off a natural route for easy scientific conversation.

Franklin's tenure at the MRC ended on an unfortunate note. As a condition to agreeing to transfer her fellowship to Birkbeck, Randall made Franklin promise not to perform additional experiments on DNA, or even to think about DNA. Moreover, Franklin was forced to leave her diffraction photograph behind at King's and to leave the work of confirming DNA's structure to Wilkins. Personality conflicts were the major source of Franklin's difficulty at King's, and her status as a woman may have made her problems worse. Even today, I and many of my women colleagues find that forcefully aggressive behavior, for example, is usually considered merely irritating when exhibited by a man but is often deemed unacceptable when demonstrated by a woman.

Box 14.1 The Evolution of Franklin's Intuition

Every scientist I interviewed, except James Watson and Max Perutz, agreed that Rosalind Franklin was a superb scientist with the sharp mind and vision needed to plan, execute, and analyze a good experiment. Questions about her capabilities have centered on her ability to make intuitive leaps when interpreting results, mainly because she seemed hesitant to do so in her DNA work. Franklin often expressed the opinion that the facts should speak for themselves. Her desire to have solid proof for her ideas before publishing them helps explain her highly successful publication record. She argued that a scientist need not be highly speculative, which gave the impression that she was incapable of speculation.

Franklin's manner and scientific approach while at King's College, London were uncannily close to those described by Frederick Dainton, Franklin's physical chemistry don when she was an undergraduate at Cambridge University. Writing to biographer Anne Sayre, Dainton says that "he was attracted by [Franklin's] directness . . . including the way she defended her point of view, something she was never loath to do." Dainton found Franklin to be a "very private person with very high personal and scientific standards, and uncompromisingly honest. . . . She once told me the facts will speak for themselves, but never fully accepted my urging that she must try to help the receptor of her messages. . . . As you point out, the logical sequential arguments meant everything to Rosalind. . . ."[13]

Francis Crick has suggested that Franklin's aversion to speculative thinking about DNA excludes her from the rank of the great scientists. He has noted, though, that the problem might not have been that Franklin lacked intuition, but rather that she might not have trusted it. Moreover, Crick has suggested that Franklin's intuition seemed to be rapidly improving as she matured as a scientist while working with tobacco mosaic virus. And if Franklin did not trust her intuition while working on DNA, it may have been at least partially because she did not have anyone at King's with whom she could properly discuss her ideas.

Box 14.2 The Helix Funeral Invitation

Maurice Wilkins insisted that DNA was helical before there was clear evidence. That greatly annoyed Rosalind Franklin, as did so many things about him. Franklin was well aware that the x-ray diffraction photographs of the A form, unlike those of the B form, lacked the clear X-shaped pattern indicative of a helix. Instead, they displayed a detailed and confusing assortment of reflections that she could only interpret through the complicated and daunting procedure of cylindrical Patterson analysis. Therefore, when, over a period of about five months starting on 18 April 1952, Franklin recorded some misleading, apparently asymmetrical double orientation data in the A form, she got perverse pleasure out of possibly annoying Wilkins with her result. As a joke, she penned the "death of the helix" funeral invitation reproduced here. That she wrote the invitation is clearly substantiated in interviews with Raymond Gosling and Wilkins and in many other professional accounts.

The extent and significance of the distribution of the invitation is another matter. During interviews, only diffraction expert Alec Stokes said that he had received one. A few people said that they saw an invitation posted on a bulletin board, and most said that they had never even seen one. That testimony is incompatible with the often repeated claim that a multitude of these cards were sent out.

Historians Horace Judson and Robert Olby argued that the invitation indicated Franklin was antihelical, although Olby notes it was a joke as well.[10,11] To the contrary: Franklin's student Raymond Gosling is adamant that Franklin considered the B form, with its striking x-ray pattern, to be helical. Examinations of Franklin's notebooks have led to the same conclusion, although it is also clear that, for a while, Franklin definitely had her doubts about the helicity of the A form.

■ Watson's view: *The Double Helix*

In *The Double Helix*, Watson bases his account of Franklin on recollections of their three brief meetings between 1951 and 1953, and on repeated complaints about her from Wilkins. The "Rosy" that Watson describes is a caricature based on the more difficult aspects of Franklin's personality. His portrayal—a far cry from the competent scientist described by her colleagues or the fascinating person described by her friends—is an effective device for promoting the idea that Watson and Crick had to rescue DNA data from—as Watson's book puts it—this "belligerent" woman who could not "keep her emotions under control" and who did not know how to interpret her own data. Watson falsely depicts Franklin as Wilkins's assistant, incapable and unworthy of Nobel Prize-caliber work. His book was published against the vehement protest of key DNA participants, who were upset about its numerous inaccuracies.[9]

Unfortunately, Watson's admittedly fascinating, irreverent, and concise book has sold millions of copies and is for many, especially in the US, the primary source of information about DNA history. That such a one-sided account both is presented as historical fact and has had tremendous influence is worrisome. Watson's treatment of Franklin, who was then deceased and not protected by libel laws, is especially troublesome.

Watson's depiction of Franklin went largely unchallenged, at least in print, until Anne Sayre published her 1975 biography *Rosalind Franklin and DNA* (W. W. Norton, 1975). Sayre asked awkward, science-related questions: Why would Franklin give her data to Watson, Crick, and Wilkins, three scientists who seemed to have nothing to offer in return at that time? Where would Watson and Crick be without Franklin's results? Why did they not acknowledge Franklin's contributions clearly and appropriately? Regrettably, Sayre's influence was diminished because she was misled about the conditions for women at the MRC. She mistakenly assumed those conditions to be as problematic as they were at the rest of King's, and therefore incorrectly interpreted Franklin's problems at King's in terms of gender issues. Subsequent accounts, though in many ways excellent, typically are dismissive of gender concerns[10] and personality issues.[11] They don't address the awkward questions Sayre raises nor do they sufficiently emphasize that the Watson and Crick model was made possible by Franklin's data. Fortunately, Brenda Maddox's outstanding and comprehensive 2002 biography of Franklin[12] does consider questions of acknowledgment and corrects misconceptions about gender and personality issues. Perhaps it will educate the public more fully than have previous texts. (See Physics Today, February 2003, page 61, for a review of Maddox's book.)

■ Inadequate Acknowledgment

In their 1953 paper, Watson and Crick state that they had been "stimulated by a knowledge of the general nature of the unpublished experimental results and ideas of Dr. M. H. F. Wilkins, Dr. R. E. Franklin and their co-workers at King's College, London." That oblique acknowledgment misrepresented Franklin's role and, whatever its intention, left most people with the impression that her work mainly served to confirm that of Watson and Crick. It has to be one of the greatest understatements in the history of scientific writing.

Given her temper, it is likely that Franklin would have been very angry if she had known the extent to which Watson and Crick used her data. In a 1951 incident, Franklin was furious that a conference acquaintance planned on publishing an idea of hers without giving proper acknowledgment. She shocked him by the tone of her letter in which she demanded coauthorship.

In 1954, Crick and Watson published a detailed methods paper in the *Proceedings of the Royal Society*. In that paper, their acknowledgment of Franklin is often ambiguous. Three of the four times they mention the importance of her data, they link it with mention of Wilkins's data first in a way that suggests the two scientists' contributions were of equal importance. Twice they follow what appears to be a clear acknowledgment of Franklin's contributions with a recanting qualifying statement, for example, "we should at the same time mention that the details of [the King's College's] X-ray photographs were not shown to us and that the formulation of the structure was largely the result of extensive model building in which the main effort was to find any structure which was stereochemically feasible." They might not have known all of the details, but they had access to a significant number of them (see also box 3 below).

Another lost opportunity for acknowledging Franklin occurred during the 1962 Nobel Prize ceremony. Neither Watson nor Crick thanked Franklin for making their discovery possible. Indeed, neither mentioned her name, although, according to Wilkins, Crick did ask him to mention Franklin. That request was a dubious shifting of responsibility given Wilkins's antipathy toward Franklin, and Wilkins in fact made only minor mention of her. Crick spoke on the genetic code, which had nothing to do with Franklin's work. Watson spoke on RNA, including RNA viruses. Franklin was an expert on those viruses, yet, in his 59 citations, Watson managed to omit any reference to her work. To this day, Watson emphasizes the opinion that Franklin, although a gifted experimentalist, could not properly interpret all of her own DNA data.

Box 14.3 Frustrated Contributors to DNA Structure

Jerry Donohue, the chemist who had insisted that James Watson try modeling DNA with the keto form of nitrogenous bases, expressed to historian Horace Judson his frustration at not being sufficiently acknowledged. "Let's face it," Donohue wrote, "if the fates hadn't ordained that I share an office with Watson and [Francis] Crick in the Cavendish [Laboratory] in 1952–3, they'd *still* be puttering around trying to pair 'like-with-like' enol forms of the bases."[14] Erwin Chargaff similarly felt insufficiently acknowledged for discovering that the DNA bases guanine and cytosine (and likewise adenine and thymine) are present in approximately equal amounts.

Maurice Wilkins was frustrated because he, understandably, had expected to be collaborating with Watson and Crick. In 1951, Rosalind Franklin joined the Medical Research Council unit at King's College, London. At that time, Wilkins lacked her years of experimental diffraction experience; like most of his MRC colleagues at the King's biophysics unit, he was an accomplished scientist struggling to apply his prior experience in physics to biological problems. As a result, other MRC units jokingly called the biophysics unit "Randall's Circus," after director John Randall. Wilkins acknowledged the nature of Randall's unit when he said of Franklin, "We were amateurs, but she was a professional."

Wilkins, though, learned quickly and acquired some valuable collaborators. By early 1953, approximately eight months after Franklin recorded her famous photograph #51, he and Herbert Wilson were able to use a variety of DNA sources to obtain x-ray photographs sufficiently detailed for structural studies.[15] Later, Wilkins designed a higher-resolution x-ray camera that enabled his team to take diffraction photographs that helped to confirm and refine the Watson and Crick structure. Wilkins had initiated DNA structural studies at King's and, in 1950 with Raymond Gosling, had taken the best DNA photograph to date—one that showed a high degree of crystallinity and thus implied that DNA structure could be solved by x-ray diffraction. Wilkins had recognized DNA's genetic importance long before Crick, obtained high-quality DNA from Rudolf Signer, and supported the hiring of experienced x-ray diffraction expert Rosalind Franklin. Wilkins is well respected by the King's staff that I interviewed and his contributions were deemed worthy of a Nobel Prize.

Franklin is prominent in virtually every telling of DNA history, but she is painted differently in various accounts. Watson and Crick made one of the most important and impressive scientific discoveries of the 20th century, but their golden helix is tarnished by the way they have treated Franklin and Wilkins. A meaningful gesture, given that it was Franklin's data that Watson and Crick most directly used, would be for scientists to refer to the "Watson, Crick, and Franklin structure for DNA."

It is important to stop demeaning Franklin's reputation, but equally important to avoid obscuring her more difficult personality traits. She should not be put on a pedestal as a symbol of the unfair treatment accorded to many women in science. Her complicated relationship with Wilkins has been treated in overly simplistic ways. Distorted accounts, which inaccurately portray the three Nobel Prize winners as well as Franklin, are unfortunate and unnecessary: There was enough glory in the work of the four to be shared by them all.

I thank Barbara Low, Mary Singleton, Martha Breed, Marilyn Goldfeather, and Paulina Miner for their many helpful editorial suggestions, and Richard Hasbrouck for his assistance in preparing figures.

References

1. J. D. Bernal, *Nature* **182**, 154 (1958).
2. S. Furberg, *Acta Crystallogr.* **3**, 325 (1950); *Acta Chem. Scand.* **6**, 634 (1952).
3. These figures are quoted in Franklin's 1952 Medical Research Council report and her unpublished 17 March 1953 draft paper. The draft was discovered many years later and written about by A. Klug in *Nature* **219**, 808 (1968); **219**, 843 (1968); **248**, 787 (1974).
4. W. T. Astbury, F. O. Bell, *Nature* **141**, 747 (1938); *Cold Spring Harb. Symp. Quant. Biol.* **6**, 112 (1938).
5. R. E. Franklin, R. G. Gosling, *Acta Crystallogr.* **6**, 673 (1953); **6**, 678 (1953).
6. J. Broomhead, *Acta Crystallogr.* **4**, 92 (1951).
7. A. Klug, *Nature* **219**, 808 (1968); **219**, 843 (1968); **248**, 787 (1974).
8. M. Bailey, *American Women in Science: A Biographical Dictionary*, ABC-CLIO, Santa Barbara, Calif. (1994); G. Kass-Simon, P. Farnes, eds., *Women of Science: Righting the Record* Indiana U. Press, Bloomington, Ind. (1990); M. B. Ogilvie, *Women in Science: Antiquity Through the Nineteenth Century. A Biographical Dictionary With Annotated Bibliography*, MIT Press, Cambridge, Mass. (1986);

M. W. Rossiter, *Women Scientists in America: Struggles and Strategies to 1940*, Johns Hopkins U. Press, Baltimore, Md. (1982); M. W. Rossiter, *Women Scientists in America: Before Affirmative Action, 1940–1972*, Johns Hopkins U. Press, Baltimore, Md. (1995).

9. See for example, W. Sullivan in J. D. Watson's *The Double Helix: A Personal Account of the Discovery of the Structure of DNA*, Norton Critical Edition, G. S. Stent, ed., W. W. Norton, New York (1980). Copies of original letters are in the Norman Collection on the History of Molecular Biology in Novato, Calif.

10. F. H. Portugal, J. S. Cohen, *A Century of DNA: A History of the Discovery of the Structure and Function of the Genetic Substance*, MIT Press, Cambridge, Mass. (1977); H. F. Judson, *The Eighth Day of Creation: Makers of the Revolution in Biology*, CSHL Press, Plainview, N.Y. (1996).

11. R. C. Olby, *The Path to the Double Helix: The Discovery of DNA*, Dover, New York (1994).

12. B. Maddox, *Rosalind Franklin: The Dark Lady of DNA*, HarperCollins, New York (2002).

13. The original letter is in the Anne Sayre Collection of the Microbiological Society Archives at the University of Maryland, Baltimore County. See also ref. 12.

14. The original 5 March 1976 letter is in the Anne Sayre Collection of the Microbiological Society Archives at the University of Maryland, Baltimore County. See also ref. 12.

15. H. R. Wilson, *Trends Biochem. Sci.* **13**, 275 (1988).

Genomics and Its Impact on Science and Society

Reading 15

■ The Human Genome Project and Beyond

A Primer

Cells are the fundamental working units of every living system. All the instructions needed to direct their activities are contained within the chemical DNA (deoxyribonucleic acid).

DNA from all organisms is made up of the same chemical and physical components. The **DNA sequence** is the particular side-by-side arrangement of bases along the DNA strand (e.g., ATTCCGGA). This order spells out the exact instructions required to create a particular organism with its own unique traits.

The **genome** is an organism's complete set of DNA. Genomes vary widely in size: The smallest known genome for a free-living organism (a bacterium) contains about 600,000 DNA base pairs, while human and mouse genomes have some 3 billion (see p. 3). Except for mature red blood cells, all human cells contain a complete genome.

DNA in each human cell is packaged into 46 **chromosomes** arranged into 23 pairs. Each chromosome is a physically separate molecule of DNA that ranges in length from about 50 million to 250 million base pairs. A few types of major chromosomal abnormalities, including missing or extra copies or gross breaks and rejoinings (translocations), can be detected by microscopic examination. Most changes in DNA, however, are more subtle and require a closer analysis of the DNA molecule to find perhaps single-base differences.

Each chromosome contains many **genes**, the basic physical and functional units of heredity. Genes are specific sequences of bases that encode instructions on how to make proteins. Genes comprise only about 2% of the human genome; the remainder consists of noncoding regions, whose functions may include providing chromosomal structural integrity and regulating where, when, and in what quantity proteins are made. The human genome is estimated to contain some 25,000 genes.

Although genes get a lot of attention, the **proteins** perform most life functions and even comprise the majority of cellular structures. Proteins are large, complex molecules made up of chains of small chemical compounds called amino acids. Chemical properties that distinguish the 20 different amino acids cause the protein chains to fold up into specific three-dimensional structures that define their particular functions in the cell.

The constellation of all proteins in a cell is called its **proteome**. Unlike the relatively unchanging genome, the dynamic proteome changes from minute to minute in response to tens of thousands of intra-and extracellular environmental signals. A protein's chemistry and behavior are determined by the gene sequence and by the number and identities of other proteins made in the same cell at the same time and with which it associates and reacts. Studies to explore protein structure and activities, known as proteomics, will be the focus of much research for decades to come and will help elucidate the molecular basis of health and disease.

■ The Human Genome Project, 1990–2003

A Brief Overview

Though surprising to many, the Human Genome Project (HGP) traces its roots to an initiative in the U.S. Department of Energy (DOE). Since 1947, DOE and its predecessor agencies have been charged by Congress with developing new energy resources and technologies and pursuing a deeper understanding of potential health and environmental risks posed by their production and use. Such studies, for example, have provided the scientific basis for individual risk assessments of nuclear medicine technologies.

Genomics and Its Impact on Science and Society. *U.S. DOE Human Genome Project, www.ornl.gov/sci/techresources/Human_Genome/publicat/primer2001/1.shtml.*

In 1986, DOE took a bold step in announcing the Human Genome Initiative, convinced that its missions would be well served by a reference human genome sequence. Shortly thereafter, DOE joined with the National Institutes of Health to develop a plan for a joint HGP that officially began in 1990. During the early years of the HGP, the Wellcome Trust, a private charitable institution in the United Kingdom, joined the effort as a major partner. Important contributions also came from other collaborators around the world, including Japan, France, Germany, and China.

Ambitious Goals

The HGP's ultimate goal was to generate a high-quality reference DNA sequence for the human genome's 3 billion base pairs and to identify all human genes. Other important goals included sequencing the genomes of model organisms to interpret human DNA, enhancing computational resources to support future research and commercial applications, exploring gene function through mouse-human comparisons, studying human variation, and training future scientists in genomics.

The powerful analytical technology and data arising from the HGP present complex ethical and policy issues for individuals and society. These challenges include privacy, fairness in use and access of genomic information, reproductive and clinical issues, and commercialization (see p. 8). Programs that identify and address these implications have been an integral part of the HGP and have become a model for bioethics programs worldwide.

A Lasting Legacy

In June 2000, to much excitement and fanfare, scientists announced the completion of the first working draft of the entire human genome. First analyses of the details appeared in the February 2001 issues of the journals *Nature* and *Science*. The high-quality reference sequence was completed in April 2003, marking the end of the Human Genome Project—2 years ahead of the original schedule. Coincidentally, it also was the 50th anniversary of Watson and Crick's publication of DNA structure that launched the era of molecular biology.

Available to researchers worldwide, the human genome reference sequence provides a magnificent and unprecedented biological resource that will serve throughout the century as a basis for research and discovery and, ultimately, myriad practical applications. The sequence already is having an impact on finding genes associated with human disease (see p. 3). Hundreds of other genome sequence projects—on microbes, plants, and animals—have been completed since the inception of the HGP, and these data now enable detailed comparisons among organisms, including humans.

Many more sequencing projects are under way or planned because of the research value of DNA sequence, the tremendous sequencing capacity now available, and continued improvements in technologies. Sequencing projects on the genomes of many microbes, as well as the chimpanzee, pig, sheep, and domestic cat, are in progress.

Beyond sequencing, growing areas of research focus on identifying important elements in the DNA sequence responsible for regulating cellular functions and providing the basis of human variation. Perhaps the most daunting challenge is to begin to understand how all the "parts" of cells—genes, proteins, and many other molecules—work together to create complex living organisms. Future analyses of this treasury of data will provide a deeper and more comprehensive understanding of the molecular processes underlying life and will have an enduring and profound impact on how we view our own place in it.

■ Insights from the Human DNA Sequence

The first panoramic views of the human genetic landscape have revealed a wealth of information and some early surprises. Much remains to be deciphered in this vast trove of information; as the consortium of HGP scientists concluded in their seminal paper, ". . . the more we learn about the human genome, the more there is to explore." A few highlights follow from the first publications analyzing the sequence.

- The human genome contains 3.2 billion chemical nucleotide base pairs (A, C, T, and G).
- The average gene consists of 3,000 base pairs, but sizes vary greatly, with the largest known human gene being dystrophin at 2.4 million base pairs.

- Functions are unknown for more than 50% of discovered genes.
- The human genome sequence is almost exactly the same (99.9%) in all people.
- About 2% of the genome encodes instructions for the synthesis of proteins.
- Repeat sequences that do not code for proteins make up at least 50% of the human genome.
- Repeat sequences are thought to have no direct functions, but they shed light on chromosome structure and dynamics. Over time, these repeats reshape the genome by rearranging it, thereby creating entirely new genes or modifying and reshuffling existing genes.
- The human genome has a much greater portion (50%) of repeat sequences than the mustard weed (11%), the worm (7%), and the fly (3%).
- Over 40% of predicted human proteins share similarity with fruit-fly or worm proteins.
- Genes appear to be concentrated in random areas along the genome, with vast expanses of noncoding DNA between.
- Chromosome 1 (the largest human chromosome) has the most genes (3,168), and Y chromosome has the fewest (344).
- Particular gene sequences have been associated with numerous diseases and disorders, including breast cancer, muscle disease, deafness, and blindness.
- Scientists have identified millions of locations where single-base DNA differences (see p. 9) occur in humans. This information promises to revolutionize the processes of finding DNA sequences associated with such common diseases as cardiovascular disease, diabetes, arthritis, and cancers.

HOW DOES THE HUMAN GENOME STACK UP?

Organism	Genome Size (Base Pairs)	Estimated Genes
Human (*Homo sapiens*)	3.2 billion	25,000
Laboratory mouse (*M. musculus*)	2.6 billion	25,000
Mustard weed (*A. thaliana*)	100 million	25,000
Roundworm (*C. elegans*)	97 million	19,000
Fruit fly (*D. melanogaster*)	137 million	13,000
Yeast (*S. cerevisiae*)	12.1 million	6,000
Bacterium (*E. coli*)	4.6 million	3,200
Human immunodeficiency virus (HIV)	9,700	9

The estimated number of human genes is only one-third as great as previously thought, although the numbers may be revised as more computational and experimental analyses are performed.

Scientists suggest that the genetic key to human complexity lies not in gene number but in how gene parts are used to build different products in a process called alternative splicing. Other underlying reasons for greater complexity are the thousands of chemical modifications made to proteins and the repertoire of regulatory mechanisms controlling these processes.

■ Managing and Using the Data

Bioinformatics Boom

Massive quantities of genomic data and high-throughput technologies are now enabling studies on a vastly larger scale than ever before. Examples include simultaneously monitoring and comparing the activity of tens of thousands of genes in cancerous and noncancerous tissue. Advanced computational tools and interdisciplinary experts are needed to capture, represent, store, integrate, distribute, and analyze the data.

Bioinformatics is the term coined for the new field that merges biology, computer science, and information technology to manage and analyze the data, with the ultimate goal of understanding and modeling living systems. Computing and information demands will continue to rise with the explosive torrent of data from large-scale studies at the molecular, cellular, and whole-organism levels.

Gene Gateway: A User-Friendly Guide to Genome, Gene, and Protein Databases

All Human Genome Project data and much related information are freely available on the web, but how do you find and use these rich resources? The Gene Gateway website provides introductory guides and step-by-step tutorials that show how to access and explore genome, gene, and protein databases used by scientists. Gene Gateway demonstrates how to gather information from different databases to gain a better understanding of the molecular biology behind life processes. The site offers a free workbook downloadable in PDF format.

Tutorials

- Identify genes associated with various genetic conditions and biological processes
- Learn about mutations that cause genetic disorders
- Browse a genome and find a gene's location on a chromosome map
- View the DNA sequence of a gene or amino acid sequence of a gene's protein product
- Visualize and modify three-dimensional representations of protein structures

Medicine and the New Genetics

Gene Testing, Pharmacogenomics, and Gene Therapy

DNA underlies almost every aspect of human health, both in function and dysfunction. Obtaining a detailed picture of how genes and other DNA sequences work together and interact with environmental factors ultimately will lead to the discovery of pathways involved in normal processes and in disease pathogenesis. Such knowledge will have a profound impact on the way disorders are diagnosed, treated, and prevented and will bring about revolutionary changes in clinical and public health practice. Some of these transformative developments are described below.

Gene Testing

DNA-based tests are among the first commercial medical applications of the new genetic discoveries. Gene tests can be used to diagnose and confirm disease, even in asymptomatic individuals; provide prognostic information about the course of disease; and, with varying degrees of accuracy, predict the risk of future disease in healthy individuals or their progeny.

Currently, several hundred genetic tests are in clinical use, with many more under development, and their numbers and varieties are expected to increase rapidly over the next decade. Most current tests detect mutations associated with rare genetic disorders that follow Mendelian inheritance patterns. These include myotonic and Duchenne muscular dystrophies, cystic fibrosis, neurofibromatosis type 1, sickle cell anemia, and Huntington's disease.

Recently, tests have been developed to detect mutations for a handful of more complex conditions such as breast, ovarian, and colon cancers. Although they have limitations, these tests sometimes are used to make risk estimates in presymptomatic individuals with a family history of the disorder. One potential benefit to these gene tests is that they could provide information to help physicians and patients manage the disease or condition more effectively. Regular colonoscopies for those having mutations associated with colon cancer, for instance, could prevent thousands of deaths each year.

Some scientific limitations are that the tests may not detect every mutation associated with a particular condition (many are as yet undiscovered), and the ones they do detect may present different risks to various people and populations. Another important consideration in gene testing is the lack of effective treatments or preventive measures for many diseases and conditions now being diagnosed or predicted.

http://genomics.energy.gov/genegateway/
Gene Gateway was created as a companion to the Human Genome Landmarks wall poster (see back page).

Knowledge about the risk of potential future disease can produce significant emotional and psychological impacts. Because genetic tests reveal information about individuals and their families, test results can affect family dynamics. Results also can pose risks for population groups if they lead to group stigmatization.

Other issues related to gene tests include their effective introduction into clinical practice, the regulation of laboratory quality assurance, the availability of testing for rare diseases, and the education of healthcare providers and patients about correct interpretation and attendant risks.

Families and individuals who have genetic disorders or are at risk for them often seek help from medical geneticists (an M.D. specialty) and genetic counselors (graduate-degree training). These professionals can diagnose and explain disorders, review available options for testing and treatment, and provide emotional support. (For more information, see the URL for Medicine and the New Genetics, p. 12.)

Pharmacogenomics: Moving Away from "One-Size-Fits-All" Therapeutics

Within the next decade, researchers will begin to correlate DNA variants with individual responses to medical treatments, identify particular subgroups of patients, and develop drugs customized for those populations. The discipline that blends pharmacology with genomic capabilities is called pharmacogenomics.

More than 100,000 people die each year from adverse responses to medications that may be beneficial to others. Another 2.2 million experience serious reactions, while others fail to respond at all. DNA variants in genes involved in drug metabolism, particularly the cytochrome P450 multigene family, are the focus of much current research in this area. Enzymes encoded by these genes are responsible for metabolizing most drugs used today, including many for treating psychiatric, neurological, and cardiovascular diseases. Enzyme function affects patient responses to both the drug and the dose. Future advances will enable rapid testing to determine the patient's genotype and guide treatment with the most effective drugs, in addition to drastically reducing adverse reactions.

Genomic data and technologies also are expected to make drug development faster, cheaper, and more effective. Most drugs today are based on about 500 molecular targets, but genomic knowledge of genes involved in diseases, disease pathways, and drug-response sites will lead to the discovery of thousands of additional targets. New drugs, aimed at specific sites in the body and at particular biochemical events leading to disease, probably will cause fewer side effects than many current medicines. Ideally, genomic drugs could be given earlier in the disease process. As knowledge becomes available to select patients most likely to benefit from a potential drug, pharmacogenomics will speed the design of clinical trials to market the drugs sooner.

Gene Therapy, Enhancement

The potential for using genes themselves to treat disease or enhance particular traits has captured the imagination of the public and the biomedical community. This largely experimental field—gene transfer or gene therapy—holds potential for treating or even curing such genetic and acquired diseases as cancers and AIDS by using normal genes to supplement or replace defective genes or to bolster a normal function such as immunity.

Almost 1,200 clinical gene-therapy trials were identified worldwide in 2006.* The majority (67%) take place in the United States, followed by Europe (29%). Although most trials focus on various types of cancer, studies also involve other multigenic and monogenic, infectious, and vascular diseases. Most current protocols are aimed at establishing the safety of gene-delivery procedures rather than effectiveness.

Gene transfer still faces many scientific obstacles before it can become a practical approach for treating disease. According to the American Society of Human Genetics' Statement on Gene Therapy, effective progress will be achieved only through continued rigorous research on the most fundamental mechanisms underlying gene delivery and gene expression in animals.

---

*Source: *Journal of Gene Medicine* website (www.wiley.co.uk/genetherapy/clinical/), August 2006.

■ Other Anticipated Benefits of Genetic Research

Expanding Impacts of New Technologies, Resources

Rapid progress in genome science and a glimpse into its potential applications have spurred observers to predict that biology will be the foremost science of the 21st Century. Technology and resources generated by the Human Genome Project and other genomic research already are having major impacts across the life sciences. The biotechnology industry employed more than 250,000 people in 2006, and revenues for 2005 totaled more than $50.7 billion.* Future revenues are expected to reach trillions of dollars.

A list of some current and potential applications of genome research follows. More studies and public discussion are required for eventual validation and implementation of some of these uses (see p. 167).

Molecular Medicine
- Improve diagnosis of disease
- Detect genetic predispositions to disease
- Create drugs based on molecular information
- Use gene therapy and control systems as drugs
- Design "custom drugs" based on individual genetic profiles

Microbial Genomics
- Rapidly detect and treat pathogens (disease-causing microbes) in clinical practice
- Develop new energy sources (biofuels)
- Monitor environments to detect pollutants
- Protect citizenry from biological and chemical warfare
- Clean up toxic waste safely and efficiently

Risk Assessment
- Evaluate the health risks faced by individuals who may be exposed to radiation (including low levels in industrial areas) and to cancer-causing chemicals and toxins

Bioarchaeology, Anthropology, Evolution, and Human Migration
- Study evolution through germline mutations in lineages
- Study migration of different population groups based on maternal genetic inheritance
- Study mutations on the Y chromosome to trace lineage and migration of males
- Compare breakpoints in the evolution of mutations with population ages and historical events

DNA Identification
- Identify potential suspects whose DNA may match evidence left at crime scenes
- Exonerate people wrongly accused of crimes
- Identify crime, catastrophe, and other victims
- Establish paternity and other family relationships
- Identify endangered and protected species as an aid to wildlife officials (e.g., to prosecute poachers)
- Detect bacteria and other organisms that could pollute air, water, soil, and food
- Match organ donors with recipients in transplant programs
- Determine pedigree for seed or livestock breeds
- Authenticate consumables such as caviar and wine

Agriculture, Livestock Breeding, and Bioprocessing
- Grow disease-, insect-, and drought-resistant crops
- Optimize crops for bioenergy production
- Breed healthier, more productive, disease-resistant farm animals

---

*Source: Biotechnology Industry Organization website (www.bio.org), June 2008.

- Grow more nutritious produce
- Develop biopesticides
- Incorporate edible vaccines into food products
- Develop new environmental cleanup uses for plants such as tobacco

Societal Concerns Arising from the New Genetics

Critical Policy and Ethical Issues

From its inception, the Human Genome Project dedicated funds to identify and address the ethical, legal, and social issues surrounding the availability of new genetic data and capabilities. Examples of such issues follow.*

- **Privacy and confidentiality of genetic information.** Who owns and controls genetic information? Is genetic privacy different from medical privacy?
- **Fairness in the use of genetic information by insurers, employers, courts, schools, adoption agencies, and the military, among others.** Who should have access to personal genetic information, and how will it be used?
- **Psychological impact, stigmatization, and discrimination due to an individual's genetic makeup.** How does personal genetic information affect self-identity and society's perceptions?
- **Reproductive issues including adequate and informed consent and the use of genetic information in reproductive decision making.** Do healthcare personnel properly counsel parents about risks and limitations? What larger societal issues are raised by new reproductive technologies?
- **Clinical issues including the education of doctors and other health-service providers, people identified with genetic conditions, and the general public; and implementation of standards and quality-control measures.** How should health professionals be prepared for the new genetics? How can the public be educated to make informed choices? How will genetic tests be evaluated and regulated for accuracy, reliability, and usefulness? (Currently, there is little regulation.) How does society balance current scientific limitations and social risk with long-term benefits?
- **Fairness in access to advanced genomic technologies.** Who will benefit? Will there be major worldwide inequities?

New Genetics Privacy Act Becomes Law

The Genetic Information Nondiscrimination Act (GINA) became law on May 21, 2008. GINA prohibits U.S. health insurance companies and employers from discrimination on the basis of information derived from genetic tests. In addition, insurers and employers are not allowed under the law to request or demand a genetic test.

- **Uncertainties associated with gene tests for susceptibilities and complex conditions (e.g., heart disease, diabetes, and Alzheimer's disease).** Should testing be performed when no treatment is available or when interpretation is unsure? Should children be tested for susceptibility to adult-onset diseases?
- **Conceptual and philosophical implications regarding human responsibility, free will vs genetic determinism, and understanding of health and disease.** Do our genes influence our behavior, and can we control it? What is considered acceptable diversity? Where is the line drawn between medical treatment and enhancement?
- **Health and environmental issues concerning genetically modified (GM) foods and microbes.** Are GM foods and other products safe for humans and the environment? How will these technologies affect developing nations' dependence on industrialized nations?
- **Commercialization of products including property rights (patents, copyrights, and trade secrets) and accessibility of data and materials.** Will patenting DNA sequences limit their accessibility and development into useful products?

---

*For more information, see the Ethical, Legal, and Social Issues URL, p. 169–170.

■ Beyond the Human Genome Project—What's Next?

Genuine Sequences: Paving the Way for a More Comprehensive Understanding

Building a "Systems Level" View of Life

DNA sequences generated in hundreds of genome projects now provide scientists with the "parts lists" containing instructions for how an organism builds, operates, maintains, and reproduces itself while responding to various environmental conditions. We still have very little knowledge of how cells use this information to "come alive," however, and the functions of most genes remain unknown. Nor do we understand how genes and the proteins they encode interact with each other and with the environment. If we are to realize the potential of the genome projects, with far-ranging applications to such diverse fields as medicine, energy, and the environment, we must obtain this new level of knowledge.

One of the greatest impacts of having whole-genome sequences and powerful new genomic technologies may be an entirely new approach to conducting biological research. In the past, researchers studied one or a few genes or proteins at a time. Because biological processes are intertwined, these strategies provided incomplete—and often inaccurate—views. Researchers now can approach questions systematically and on a much grander scale. They can study all the genes expressed in a particular environment or all the gene products in a specific tissue, organ, or tumor. Other analyses will focus on how tens of thousands of genes and proteins work together in interconnected networks to orchestrate the chemistry of life. These holistic studies are the focus of a new field called "systems biology" (see DOE Genomics:GTL Program, p. 168).

Charting Human Variation

Slight variations in our DNA sequences can have a major impact on whether or not we develop a disease and on our responses to such environmental factors as infectious microbes, toxins, and drugs. One of the most common types of sequence variation is the single nucleotide polymorphism (SNP). SNPs are sites in a genome where individuals differ in their DNA sequence, often by a single base. For example, one person might have the DNA base A where another might have C, and so on. Scientists believe the human genome has at least 10 million SNPs, and they are generating different types of maps of these sites, which can occur in both genes and noncoding regions.

Sets of SNPs on the same chromosome are inherited in blocks (haplotypes). In 2005 a consortium of researchers from six countries completed the first phase of a map of SNP patterns that occur across populations in Africa, Asia, and the United States. Researchers hope that dramatically decreasing the number of individual SNPs to be scanned will provide a shortcut for tracking down the DNA regions associated with such common complex diseases as cancer, heart disease, diabetes, and some forms of mental illness. The new map also may be useful in understanding how genetic variation contributes to responses to environmental factors.

■ DOE Genomics:GTL Program

Exploring Genomes for Energy and Environmental Applications

The Genomics:GTL (formerly Genomes to Life) program of the U.S. Department of Energy (DOE) is using the Human Genome Project's technological achievements to help solve our growing energy and environmental challenges.

Today, genomics is the starting point for a new level of exploration across the life sciences. The GTL research program uses genomic (DNA) sequences of microbes and plants to launch large-scale investigations into their wide-ranging biochemical capabilities having potential applications in bioenergy and the environment (see sidebar below). Before these biological processes can be safely and economically harnessed for such uses, however, they must be understood in far greater detail and in the context of their operations within a dynamic, living organism.

To obtain this whole-systems knowledge, GTL investigates relevant plant and microbial properties on multiple levels. Starting with the DNA sequence, studies follow its expression (e.g., protein production, interactions, and regulation) in individual cells and populations of cells or organisms in ecosystems. Integrating genomic and many other data types into a computerized knowledgebase will stimulate new research strategies and insights needed for specialized applications.

GTL Investigations of Microbial and Plant Genomes

Microbes and plants have evolved unique biochemistries, offering a rich resource that can be applied to diverse national needs. Some recent projects funded by the DOE Genomics:GTL program highlight the potential wealth of natural capabilities available.

Plants for Biomass, Carbon Storage

Understanding the genes and regulatory mechanisms controlling growth and other traits in the recently sequenced poplar tree may lead to its use for bioethanol production and for sequestration (storage) of carbon.

Microbes Living in Termites: A Potential Source of Enzymes for Bioenergy Production

GTL researchers are investigating bacteria that live in termite hindguts and churn out wood-digesting enzymes. These proteins may be usable for breaking down plant cellulose into sugars needed for ethanol production. Termites also produce hydrogen as a by-product, a process that potentially could be reproduced on a larger scale.

Synthetic Nanostructures: Harnessing Microbial Enzyme Functions

Enzymes incorporated into synthetic membranes can carry out some of the functions of living cells and may be useful for generating energy, inactivating contaminants, and sequestering atmospheric carbon.

Genomes for Bioenergy

Cellulosic Biomass: An Abundant, Secure Energy Source to Reduce U.S. Dependence on Gasoline

Bioethanol made from cellulosic biomass—the inedible, fibrous portions of plants—offers a renewable, sustainable, and expandable domestic resource to meet the growing demand for transportation fuels and reduce our dependence on oil.

The United States now produces 7 billion gallons of corn-grain ethanol per year, a fraction of the 142 billion gallons of transportation fuel used annually. Cellulosic ethanol has the potential to dramatically increase the availability of ethanol and help meet the national goal of displacing 30% of gasoline by 2030.

Cellulose is the most abundant biological material on earth. The crops used to make cellulosic ethanol (e.g., postharvest corn plants—not corn grain—and switchgrass) can be grown in most states and often on marginal lands. As with ethanol from corn grain, cellulose-based ethanol can be used as a fuel additive to improve gasoline combustion in today's vehicles. Modest engine modifications are required to use higher blends (85% ethanol). Additionally, the amount of carbon dioxide emitted to the atmosphere from producing and burning ethanol is far less than that released from gasoline.

To accelerate technological breakthroughs, the DOE Genomics:GTL program will establish research centers to target specific DOE mission challenges. Three DOE Bioenergy Research Centers are focused on overcoming biological challenges to cellulosic ethanol production. In addition to ethanol, these centers are exploring ways to produce a new generation of petroleum-like biofuels and other advanced energy products from cellulosic biomass.

For More Information

Related Websites

- **Human Genome Project Information**
  www.ornl.gov/hgmis/home.shtml
- **Medicine and the New Genetics**
  www.ornl.gov/hgmis/medicine/
- **Ethical, Legal, and Social Issues**
  www.ornl.gov/hgmis/elsi/
- **Genetics Privacy and Legislation**
  www.ornl.gov/hgmis/elsi/legislat.shtml

- **Gene Gateway**
  genomics.energy.gov/genegateway/
- **Image Gallery (downloadable)**
  genomics.energy.gov/gallery/
- **Resources for Teachers**
  www.ornl.gov/hgmis/education/
- **Resources for Students**
  www.ornl.gov/hgmis/education/students.shtml
- **Careers in Genetics and the Biosciences**
  www.ornl.gov/hgmis/education/careers.shtml
- **DOE Joint Genome Institute**
  www.jgi.doe.gov
- **NIH National Human Genome Research Institute**
  www.genome.gov
- **Genomes Online Database (GOLD)**
  www.genomesonline.org
- **National Center for Biotechnology Information**
  www.ncbi.nlm.nih.gov
- **DOE Genomics:GTL Program**
  genomicsgtl.energy.gov
- **Focus on Biofuels**
  genomicsgtl.energy.gov/biofuels/
- **National Geographic: The Genographic Project**
  www.nationalgeographic.com/genographic/
- **International HapMap Project**
  www.hapmap.org

Free Wall Poster of Human Chromosomes and Genes

The poster also features sidebars explaining genetic terms, with URLs for finding more detailed information (see web companion. Gene Gateway, p. 4).

Order free copies via the web:
- genomics.energy.gov/posters/
  Or see contact information below left.

Genomics and Its Impact on Science and Society: The Human Genome Project and Beyond

This document was revised in June 2008 by the Genome Management Information System (GMIS) at Oak Ridge National Laboratory, Oak Ridge, Tennessee, for the Office of Biological and Environmental Research within the U.S. Department of Energy Office of Science.

Free Copies and Presentation Materials
- **Individual print copies or class sets**
  Contact: genome@ornl.gov, 865/574-0597
- **Downloadable files**
  A PDF version of this document and accompanying PowerPoint file are accessible via the web (www.ornl .gov/hgmis/publicat/primer).

Ecology and Environmentalism

Reading 16

Peter J. Bowler & Iwan Rhys Morus

At first sight it might seem obvious that the two topics listed in the title above should be linked together. The environmentalist movement has sought to warn of the dangers posed by humanity's evermore powerful efforts to exploit the world and its inhabitants through industry and intensive agriculture. It points to the increasingly common catastrophes that can be attributed to the uncontrolled exploitation of the world's resources and notes that we are now witnessing a mass extinction of geological proportions caused by the destruction of species' natural habitats. If we are not careful, the environmentalists warn, we shall wipe ourselves out by rendering the whole world uninhabitable. To make this point they sometimes call on the science of ecology, which seeks to describe and understand the relationships between organisms and their environment. Indeed the term "ecological" is often taken to mean "environmentally beneficial," as though the science went hand in hand with the social philosophy that seeks to defend the natural world (see the title of Bramwell's 1989 book, which is actually about environmentalism). Many assume that ecology is a science created by environmentalists to provide them with the information they need about the balance of nature and the ways in which disturbing influences such as human exploitation upset and ultimately destroy that balance. Such an interpretation of the origins of ecology would take it for granted that the science is based on a holistic worldview that seeks to understand how everything in nature interacts to produce a harmonious and self-sustaining whole. Ecology is the science behind James Lovelock's image of the earth as "Gaia"—a sustaining mother to all living things who will not hesitate to discipline one of her children if it gets out of line and threatens the whole.

One pioneering study by Donald Worster (1985) sought to present such a unified picture of the origins of both environmentalist thought and scientific ecology. But subsequent work has uncovered a more complex and far less coherent pattern of relationships. To a large extent, the environmentalist movement has opposed modern science as the handmiden of industrialization, seeking its image of nature in a romantic impresionism rather than in scientific analysis. To the extent that it has had an impact on science, it has done so by encouraging a holistic methodology that openly challenges the materialistic and reductionist approach favored by the majority of scientists. There are thus some forms of scientific ecology that do draw inspiration from environmentalist concerns—but there are others that owe their origins to the reductionist viewpoint that is anathema to the romantic vision of natural harmony. Many of the first professional ecologists used physiology as a model, arguing that just as the physiologists saw the body as a machine, so they should apply a purely naturalistic methodology to studying how the body interacted with its environment. Some schools of ecology have remained resolutely materialistic, depicting natural relationships in terms more of a Darwinian struggle for existence than of harmony. Ecologists from these backgrounds are among the leading critics of Lovelock's efforts to depict nature as a purposeful whole that seeks to maintain the earth as an abode for life.

Modern historical studies force us to see ecology as a complex science with many historical roots. Indeed, it is not really a unified branch of science at all, since its various schools of thought have such different origins that they still find it hard to communicate with one another. Providing hard evidence for the environmentalist campaign is certainly not on most scientific ecologists' agenda. As in so many other areas, a historical study forces us to contextualize the rise of science, breaking down the more obvious links, such as those assumed to exist between ecology, holism, and environmentalism. Instead, we see the science emerging from a number of different research programs instituted in different places and times and for different purposes, some of them designed more to encourage the exploitation of the environment than to promote its protection. Far from

"Ecology and Environmentalism," from Making Modern Science: A Historical Survey *by Peter J. Bowler and Iwan Rhys Morus, pp. 213–235, 2005 University of Chicago Press. Reprinted by permission of The University of Chicago Press, publisher.*

originating as a unified response to a single philosophical message, ecology is a composite of many rival approaches that even today have not coalesced into a single discipline with a coherent methodology.

We begin with an overview of how science became associated with the drive to exploit the world's resources, then move on to an account of how the environmentalist movement emerged to counter this program. The second half of this chapter then outlines the emergence of scientific ecology from the late nineteenth century onward, showing how different research problems and different philosophical an ideological agendas promoted theoretical disagreement almost from the beginning.

Science and the Exploitation of Resources

From the Scientific Revolution of the seventeenth century onward, the rise of science has been linked to the hope that better knowledge of the world would allow a more effective use of natural resources. The ideology promoted by Francis Bacon stressed the use of observation and experimentation to build up practical knowledge that could be applied through improvements to industry and agriculture. The world was depicted as a passive source of raw materials to be exploited by humanity for its own benefit. Even the methodology of science stressed the dominance of humanity and the passivity of the natural world: the experimenter sought to isolate particular phenomena so they could be manipulated at will. There was no expectation that everything might interact in a way that would negate the insights gained from the study of the particular. If the whole universe was just a machine, there was no reason why humanity should not tinker with individual parts for its own benefit. Carolyn Merchant (1980) sees this attitude as characteristic of an increasingly "masculine" attitude toward nature (see chap. 21, "Science and Gender"). By the end of the eighteenth century, this attitude was already bearing fruit as the Industrial Revolution got underway, and in the course of the following century the role that science could play in promoting technological development became obvious to all (see chap. 17, "Science and Technology").

At the same time, science was increasingly involved in the effort to locate and exploit natural resources around the world. The voyages of discovery undertaken by navigators such as Captain James Cook were intended to bring back information on the plants and animals of remote regions for Europeans to study and classify, but they were also intended to locate new territories that might be colonized. Sir Joseph Banks accompanied Cook on his first voyage to the South Seas (1768–71) as a naturalist. In his later capacity as president of the Royal Society, he helped to coordinate the British navy's efforts to explore and map the world, often with a view to discovering useful natural resources (MacKay 1985). The voyage of H.M.S. *Beagle,* which provided Darwin with crucial insights, was undertaken to map the coast of South America, a region vital to British trade. In the 1870s, the British navy provided a vessel, H.M.S. *Challenger,* for the first deep-sea oceanographic expedition. Although much information of scientific interest was generated, funding for marine science was increasingly provided in the expectation that there would be benefits for navigation, fisheries, and other practical concerns.

On land, too, there were many expeditions designed to explore remote regions to satisfy curiosity about the world (see below), but there were also explicit signs of science's growing involvement with imperialism. Many European nations established botanical gardens at home and in their colonies with the deliberate intention of identifying commercially useful plant species and studying how foreign species could be imported as new cash crops. Kew Gardens in London was the center of the British effort, under the direction of botanists such as Joseph Dalton Hooker, a leading supporter of Darwin (Brockway 1979). The cinchona plant, source of the anti-malarial drug quinine and hence vital to European efforts to colonize the tropics, was transported via Kew from its home in South America to found commercial plantations in India. The rubber plant was smuggled out of Brazil despite a government prohibition to create the worldwide rubber-production industry. North America was transformed as European farming methods were adapted to its wide range of different environments. By the early twentieth century, the Bureau of Biological Survey under C. Hart Merriam was coordinating deliberate attempts to eradicate native "pests" such as the prairie dog that destroyed the farmers' crops. Europeans and Americans were now interfering on an unprecedented scale with natural ecosystems, destroying native habitats and importing alien species as cash crops (for a survey of these developments, see Bowler [1992]).

The Rise of Environmentalism

These developments were not without their critics, and gradually an articulate movement evolved to criticize the unrestricted exploitation—and often the consequent destruction—of the natural environment (McCormick

1989). The Romantic thinkers of the early nineteenth century celebrated the wilderness as a source of spiritual renewal and hated the industrialists who destroyed it for profit. Significantly, writers such as William Blake saw mechanistic science as a key component of the unrestrained exploitation of the natural world. A later generation of writers such as Henry Thoreau also celebrated the recuperative value of wilderness for a humanity increasingly alienated by an urban and industrialized lifestyle. In 1864, the American diplomat George Perkins Marsh wrote his *Man and Nature* to protest against the destruction of the natural environment. He warned that, contrary to early optimistic expectations, there was a degree of human destructiveness that nature might never be able to repair: "The earth is fast becoming an unfit home for its noblest inhabitant, and another era of equal human crime and human improvidence . . . would reduce it to such a condition of impoverished productiveness, of shattered surface, of climatic excess, as to threaten the depravation, barbarism and perhaps even extinction of the species" (Marsh 1965, 43). Marsh was not calling for a halt to all human interference but for better management that would allow the earth to retain its self-sustaining capacities. Partly as a result of his efforts, the U.S. government set up the Forestry Commission to manage the nation's resources, and eventually areas of woodland were set aside to be protected from logging. Public concern also led to the designation of areas of outstanding natural beauty as national parks, Yosemite Valley in California in 1864 and Yellowstone in Wyoming in 1872. The Sierra Club, founded in 1892 by John Muir, was dedicated to the protection of wilderness areas. In Europe, where there was little true wilderness left to protect, efforts were nevertheless made to create nature reserves where stable environments that had existed for centuries could be conserved (on nature reserves in Britain, see Sheal [1976]).

There was considerable tension between those who called for a more careful management of nature in order to allow resources to be renewed and an increasingly vocal movement that depicted all human interference as evil and potentially damaging to the earth as a whole. The former group was willing to call in science, in the form of the newly developed ecology, to help better understand the ways in which natural ecosystems would respond to human interference. But a more extreme form of environmentalism developed out of an alternative, more romantic vision of nature that, if it had any use for science at all, insisted that it must be a science based on holistic rather than mechanistic principles. This movement cut across all traditional political divisions and was by no means always sympathetic to a democratic approach to government. After all, the common people may well vote for more industrialization out of a short-sighted desire for more material goods. In Germany, a "religion of nature" often linked to the philosophy of the evolutionist Ernst Haeckel, became part of Nazi ideology—and the Nazis created nature reserves on ground cleared of Jews and Poles sent to the death camps. Soviet Russia had a strong environmentalist policy until Stalin's drive for industrialization led to unrestricted exploitation of the country's resources (on European environmentalism, see Bramwell [1989]).

In America, there were debates between those who saw the "dust bowl" on the Great Plains in the 1930s as part of a natural climatic cycle and those who insisted that it was a consequence of the unsuitability of the prairies for farming. The latter position was increasingly typical of the more active environmentalist movement, which allied itself with those who saw the preservation of wilderness as essential for human psychological health, to say nothing of the health of the planet as a whole. In America, Aldo Leopold's *Sand County Almanac,* published posthumously in 1949, recorded the transition of a Wisconsin game manager into an environmentalist with an emotional and aesthetic attachment to wilderness. For Leopold, scientific ecology was not enough because it needed to be supplemented by an ethical commitment that recognized that all species have a right to exist, a right that should not be compromised by human expediency: "Conservation is getting nowhere because it is incompatible with our Abrahamic concept of land. We abuse land because we regard it as a commodity belonging to us. When we see land as a community to which we belong, we may begin to use it with love and respect. There is no other way for land to survive the impact of mechanized man, nor for us to reap from it the esthetic harvest it is capable, under science, of contributing to culture" (Leopold 1966, x). Leopold's environmentalism did not rule out a role for the scientific study of nature, but that had to take place within a framework in which humanity was part of nature, not dominant over it.

Such an attitude has grown in influence, as more people have become aware of the dangers of the unrestricted exploitation of the environment. Rachel Carson's *Silent Spring* of 1962 highlighted the damage done to many species by the use of insecticides. Numerous environmental catastrophes have driven home the same message, although there are still significant differences between the ways in which different communities have responded. In America, despite the activities of those who cherish the wilderness, the public seems content to let corporate agriculture manipulate nature in the interests of producing cheaper food. In Europe, by contrast, the use of chemical fertilizers and insecticides has become unpopular, while genetic manipulation of food crops is restricted. In the Third World, however, genetic engineering is seen as perhaps the lesser of the two evils, since it might increase yields without leaving farmers dependent on expensive and potentially dangerous chemicals.

■ The Origins of Ecology

A distinct science of ecology only began to emerge at the end of the nineteenth century, although concepts we associate with the discipline had long been recognized. The Swedish naturalist Linnaeus wrote of the "balance of nature" in the mid-eighteenth century, noting that it one species increased its numbers due to favorable conditions, its predators would also increase and tend to restore the equilibrium. For Linnaeus, this was all part of God's plan of creation, and the natural theologians routinely described the adaptation of species to their physical and biological environment as an illustration of divine benevolence.

Systematic study of such relationships was also part of Alexander von Humboldt's project for a coordinated science of the natural world, which focused especially on the geographical factors that shaped different environments. Humboldt was impressed by the Romantic movement popular in the arts around 1800, with its emphasis on the ability of wilderness to inspire human emotions, but he insisted that a serious study of the natural world must use the scientific techniques of measurement and rational coordination. His aim was a science that focused on material interactions but interpreted them as parts of a coordinated whole in which each natural phenomenon was interlinked with all the others. He spent the years 1799–1804 exploring South and Central America, taking numerous scientific measurements in a variety of environments that were used to throw light on the interactions between their geological structure, physical conditions, and biological inhabitants. Humboldt made important contributions to geology—he was a follower of A. G. Werner and named the Jurassic system of rocks after the Jura Mountains of Switzerland (see chap. 5, "The Age of the Earth"). He also produced maps showing the variations of temperature and other climatic factors on a worldwide scale and others showing cross sections of mountainous regions illustrating how the characteristic vegetation changed with altitude. Humboldt's accounts of his South American voyage inspired many European scientists, including Darwin, and his emphasis on the earth as an integrated whole encouraged a whole generation to undertake systematic surveys of a variety of physical and biological phenomena. Under the influence of "Humboldtian science" biologists were taught to think in what we would now call ecological terms, looking for the ways in which the distribution of animals and plants was determined by the character of the soil and underlying rocks, the local climate, and the other native inhabitants of the region.

In the next generation, Darwinism, too, stressed the adaptation of the species to its environment but encouraged a more materialistic view of each population in competition not just with its predators but also with rivals seeking to exploit the same resources (see chap. 6, "The Darwinian Revolution"). Darwin also focused attention on biogeography, which illustrated how species adapted to new environments. It was the German Darwinist Ernst Haeckel who coined the term "oecology" in 1866 from the Greek *oikos,* referring to the operations of the family household—the ecology of a region showed how the species there interacted to exploit its natural resources. But unlike Darwin, Haeckel adopted a nonmaterialistic view of nature in which living things were active agents within a unified and progressive world. The tension between the materialistic and holistic worldviews ensured that the science of ecology would he driven by theoretical disagreements from its inception. There were a number of different research programs, each trying to tackle the complex relationships between species and their environment in a different way. Because they began different origins, they often adopted different theoretical outlooks.

The stimulus for the creation of the new biological discipline that would adopt the name ecology came from the breakdown of the descriptive or morphological approach to nature at the end of the nineteenth century. At that juncture, the emphasis was on experimentation, with physiology as the model, and a number of new biological disciplines arose in response to this challenge, including genetics. It was much harder to apply the experimental method to the study of how species relate to their environment, but there were several avenues that pointed the way to a more scientific approach to this topic. One was the increasing refinement of Humboldt's biogeographical techniques. In America, C. Hart Merriam of the Bureau of Biological Survey developed detailed maps showing the various "life zones" or habitats stretching from east to west across the continent. In 1896, Oscar Drude of the Dresden botanical garden published a fine-grained plant geography of Germany that showed how local factors such as rivers and hills shaped the vegetation of each region.

Plant physiology provided the model for other pioneers of plant ecology. Experimental studies had produced a much better understanding of how the internal functions of a plant operate, but by the end of the century a number of botanists began to realize that it would also be necessary to look at how the plant's physical environment affected these functions. This insight was especially obvious to those who worked in botanical gardens established in the tropics and other extreme environments, where the role of adaptation was crucial (Cittadino 1991). The founder of plant ecology, botanist Eugenius Warming, was trained in plant physiology in Denmark and had worked for a time in Brazil. He developed his approach as an alternative both to pure physiology and to the traditional focus of most botanists on classification (Coleman 1986). His *Plantesamfund,* published in 1895, was translated into German the following year and into English as *Oecology of Plants* in 1909. Warming could see how the physical conditions of an area determined which plants could live there, but he also realized that there was a network of interactions between the plants that were characteristic of a particular environment. These typical plants formed a natural community, each dependent in various ways on the others. The concept of a natural community had already been described by naturalists such as Stephen A. Forbes of Illinois, whose 1887 address to the Peoria Scientific Association, "The Lake as a Microcosm," had stressed that all the species inhabiting a lake were dependent on one another. It was a concept that was all too easily taken up by the opponents of materialism to argue that the community formed a kind of superorganism with a life and purpose of its own. But Warming resolutely opposed this almost mystical view of the community; for him the relationships were just a natural consequence of evolution adapting species to the biological as well as the physical environment. He acknowledged that all the species were competing with each other in a constant struggle for existence and that when the original community was disturbed (as by human interference) there was no guarantee that the original collection of species would reestablish itself. If we cut down a forest, the trees may never get a chance to grow again because the soil has been modified in a way that prevents them from reseeding themselves. This view was also characteristic of one of the first American schools of ecology founded at the University of Chicago by Henry C. Cowles.

There was another American research tradition, however, that developed around a very different viewpoint. At the state university in Nebraska, Frederic E. Clements sought to put the study of grassland ecology on a more scientific footing (Tobey 1981). The European techniques were not suited to the vast uniform areas of the prairies, and Clements realized that in these conditions the only way to get really accurate information about the plant population was iterally to count every single plant growing in a series of sample areas. He marked out measured squares or quadrats spread over a wide region and compounded the information to give a much more precise assessment of the overall population. By clearing quadrats of all vegetation, he was able to see how the natural plant community reestablished itself and became convinced that in these circumstances there was a definite sequence by which the natural or "climax" population was built up. Clements's *Research Methods in Ecology* (1905) publicized the new techniques, and the school of grassland ecology established itself, especially in institutions dealing with the practical problems of the farmers whose activities inevitably destroyed the natural climax grassland of the prairies. Clements was an influential writer and he promoted a philosophy of ecology that was very different from the materialistic approach of Warming and Cowles. He saw the natural climax population of a region in almost mystical terms: nature was predestined to move toward this community whenever it was disturbed, and the community had a reality of its own that required it to be seen as something more than a collection of competing species. Here was an ecology that seemed to derive from the romantic image of nature as a purposeful whole that resisted human interference, yet it was being used to give advice to the farmers whose activities had destroyed the natural environment of the plains.

■ Consolidation and Conflict

In the early decades of the twentieth century, the rival approaches to ecology pioneered by Warming and Clements gained enough attention for the area as a whole to become recognized as an important branch of science. But new developments continued the original tensions, and there was competition among the different research schools for control of its journals and societies and for access to government and university departments where it might flourish. In fact, despite a promising start, expansion was slow until after World War II. The British Ecological Society was the first ecological society to be founded, in 1913 (Sheal 1987), followed two years later by the Ecological Society of America (whose journal, *Ecology,* first appeared in 1920). But the new discipline's bid to establish itself in academic departments was slow, except in America, and even here the membership of the Ecological Society remained static through the interwar years. In Britain, pioneer ecologists such as Arthur G. Tansley had to struggle for academic recognition; Tansley spent some time as a Freudian psychologist and blamed the slow growth of ecology in part on the loss of promising young scientists in World War I.

In America, Clements's school of grassland ecology continued to flourish into the 1930s, when it provided support for the claim that the prairies should be returned to their natural climax of grassland to recover from the erosion of the Dust Bowl. The idealist notion of the climax community as a superorganism with a life of its own was linked by his student John Phillips to the holistic philosophy being popularized by the South African statesman Jan Christiaan Smuts, whose *Holism and Evolution* appeared in 1926. Smuts made an emotional appeal to a vision of nature as a creative process with inbuilt spiritual values and depicted evolution as a process designed to bring about complex entities whose properties were of a higher level than anything visible in their individual parts. In Britain, Tansley had to compete with South African ecologists wedded to Smuts's philosophy who were threatening to dominate ecology throughout the British Empire (Anker 2001).

Although Clements and his supporters tried to explain the Dust Bowl, the fact that the soil had disappeared effectively undermined their claim that the natural climax vegetation could reestablish itself. Other schools of ecology developed, especially in university departments that did not have to deal with the problems of the prairie farmers. Henry Allan Gleason and James C. Malin both challenged Clements's ideas by arguing that changes could take place in the vegetation of a region due to fluctuations in the climate and the natural invasion of species from other regions. In Britain, Tansley—who eventually gained a chair at Oxford—argued strenuously against Phillips's use of the superorganism concept, openly dismissing it as little more than mysticism. Yet Tansley used research methods very similar to those of the Clements's school, and it was he who coined the term "ecosystem" in 1935 to denote the system of interactions holding the species of a particular area together. For any European biologist, it seemed obvious that most apparently "natural" communities were to some extent the product of human activity, perhaps extended over centuries, so there was little point in trying to claim that a particular ecosystem had some sort of prior claim to be recognized as the only one appropriate for a certain area. Tansley and other critics also worried that promoting the idea of a superorganism would play into the hands of mystics who wanted to block any scientific study of the natural world. In continental Europe, an entirely different form of ecology based on the precise classification of all the plants in an area was developed, and in this the notion of a superorganism was simply irrelevant.

A clear indication of the fragmentary origins of ecology can be seen in the fact that it was not until the 1920s that systematic study of animal ecology began. But here, too, the tensions between the materialistic and holistic viewpoints immediately asserted themselves. At the University of Chicago, Victor E. Shelford applied Clements's approach to the study of animal communities and their dependence on the local vegetation. Also at Chicago, Warder Clyde Allee began to study animal communities on the assumption that cooperation between the members of the population is an integral part of how a species deals with its environment. Allee dismissed the Darwinian view of individual competition as the driving force of behavior and of evolution—he explicitly rejected the notion of a "pecking order" determining individuals' rank within the group. For him, evolution promoted cooperation, not competition, a view closely allied with the holistic philosophy characteristic of Clements's group. Allee and his followers also developed the political implications of their vision of natural relationships as an alternative to the "social Darwinism" that presented individual competition as natural and inevitable (Mitman 1992).

A very different approach was developed in Britain by Charles Elton, who worked at the Bureau of Animal Populations at Oxford from 1932 (Crowcroft 1991). His book *Animal Ecology* (1927) established itself as a textbook for the field and popularized the term "niche" to denote the particular way in which a species interacted with its environment. Elton had worked with the records of the Hudson's Bay Company that gave details of fluctuations in the numbers of fur-bearing animals trapped over many years. These revealed occasional massive increases in numbers (plagues of lemmings are the classic example) caused when rapidly reproducing species outstrip their natural predators in a time of plentiful resources. The occurrence of such episodes made nonsense out of the old idea of a "balance of nature" and confirmed Darwin's Malthusian image of populations constantly tending to expand to the limit of the available resources.

Elton made common cause with Tansley and with the young Julian Huxley to promote their vision of ecology, which Huxley was also concerned to link with the new Darwinism emerging in evolution theory. By denying the existence of a natural ecosystem characteristic of any environment, their approach made it easier to see the natural world as something that could be adjusted to human activity through scientific planning. Such a vision had clear social implications and was popularized in the science fiction novels being written by H. G. Wells (who also collaborated with Huxley on a major popular work, *The Science of Life,* in 1931). At this point, however, they did not envisage ecology as a subject that could be analyzed using mathematical models, partly because the rapid fluctuations in population density observed by Elton seemed unpredictable. But others were becoming more interested in the possibility of using mathematics, perhaps by seeing an analogy between the behavior of individual molecules in a gas and of individual animals interacting with their environment. The American physical chemist Alfred J. Lotka published a book on this topic in 1925, and this approach was subsequently taken up by the Italian mathematical physicist Vico Volterra, who had become interested in predicting the fluctuations in commercial fish populations. In the late 1930s, the Russian biologist G. F. Gause performed experiments on protozoa to test the "Lotka-Volterra equations," and his efforts to substantiate the mathematical techniques would play a vital role in stimulating the expansion of ecology after World War II (Kingsland 1985). For the time being, however, there were many who shared Elton's suspicions, feeling that the unpredictable dynamics of natural population changes were an unsuitable field for the application of abstract mathematical models.

■ Modern Ecology

Ecology expanded rapidly in the 1950s and 1960s as the world became more aware of the pressing environmental problems created by human activity. But the pressure was not necessarily coming from environmentalist groups. Those who sought to control and exploit nature also wanted information that would help them manage the ever more complex problems that they were confronting (Bocking 1997). The ecologists exploited the new image of a more "scientific" approach made possible by the mathematical techniques developed by Lotka and Volterra before the war. They were also able to make common cause with the Darwinian synthesis now beginning to dominate evolutionary biology following the emergence of the genetical theory of natural selection (itself based on mathematical modeling of populations). A school of population ecology emerged based on the exploitation of the Darwinian idea that competition was the driving force of natural relationships. There was no overall theoretical consensus, however, because at the same time a rival school of systems ecology emerged, exploiting analogies between ecological relationships and the stable economic structures existing in human society. Here there was a renewed focus on the harmonious nature of communities, drawing not on the old vitalistic philosophy but on the models of purposeful natural systems created in cybernetics. When James Lovelock's Gaia theory extended this approach into something that looked like the old mysticism, he was violently criticized by most biologists for abandoning the materialist ethos of science and pandering to the romanticized image of nature favored by the extreme environmentalists.

The Lotka-Volterra equations reinforced the lessons of Darwinism by implying that in a world dominated by competition, the best-adapted species in any environment would drive all rivals to extinction. This became known as the "principle of competitive exclusion," which states that there can be only one species occupying a particular niche in a particular location. This principle was tested by David Lack, a student of Julian Huxley, in the case of "Darwin's finches" on the Galapagos Islands. Although Darwin had used these birds as a classic example of specialization, later studies had shown that there were often several different species feeding in apparently the same way on the same island. Lack showed that this was not the case because each species was actually exploiting a different way of feeding—just because they were all mingling together did not mean they were taking the same food in the same way. His book *Darwin's Finches* (1947) helped to establish the new Darwinian synthesis in evolutionism and the principle of competitive exclusion in ecology, while at the same time renewing interest in Darwin's role as the founder of the selection theory.

The British-trained ecologist G. Evelyn Hutchinson, who had moved to America in 1928, launched an attack on Elton's refusal to use mathematical models in animal ecology. He argued that where there were difficulties in applying the Lotka-Volterra equations, the best approach was to modify the mathematical models, not reject the technique altogether. Hutchinson wanted to use the mathematical models to unify ecology and evolution theory, as proclaimed in the title of his 1965 book *The Ecological Theatre and the Evolutionary Play*. His student Robert MacArthur went on to found a new science of community ecology based on Darwinian principles of struggle and competitive exclusion (Collins 1986; Palladino 1991). MacArthur used mathematical models to address questions such as how close the niches could be in a particular environment and whether the niches evolved along with the species. Like Lack, MacArthur became interested in the problems posed by the structure of populations on isolated islands. He teamed up with Edward O. Wilson to develop a theory that predicted that the diversity of species on an oceanic island was directly proportional to its area. The number of species was maintained by a balance between immigration and extinction, the latter always a threat to small isolated populations. Wilson became interested in the way in which different reproductive strategies would help or hinder a species trying to establish itself on a new island and subsequently went on to develop the science of sociobiology.

Hutchinson had other interests, however, and these helped to create a rival school of systems ecology based on very different theoretical principles. He wanted to study communities using not an organismic analogy but an economic one, which traced the flow of energy and resources through the system and sought to identify feedback loops that maintained the stability of the whole. This was an approach pioneered by the Russian earth scientist V. Vernadskii, who had coined the term "biosphere" earlier in the century. The concept of feedback loops was central to the new science of cybernetics founded by Norbert Weiner to explain the activity of self-regulating machines. Hutchinson imagined such feedback loops working on a global scale to maintain the various ecosystems in a stable state. He also saw an analogy between this model of nature and the economists' attempts to depict human society as a stable system based on the cooperative use of resources. Hutchinson's student Raymond Lindemann wrote an influential paper in 1942 analyzing the flow of energy derived from the sun through the ecosystem of Cedar Bog Lake in Minnesota. This model of energy flow was then built on by the brothers Howard and Eugene Odum, the founders of systems ecology. The Odums studied the energy and resource circulations in a wide variety of environments, basing their work on the assumption that large-scale ecosystems would have a substantial robustness in the face of external threats. Some of their studies were funded by the U.S. Atomic Energy Commission, anxious about the potential damage that might be caused by nuclear war or accident. Systems ecology saw the human economy as just one aspect of a global network of energy and resource consumption and presented models suggesting that all levels of the process could be managed successfully if the flow patterns could be understood. Howard Odum's *Environment, Power and Society* (1971) presented a technocrat's dream of a society carefully structured and managed so that it could maintain itself even in the face of the more restricted levels of resources that will be available to humanity in the future (Taylor 1988).

Community ecology and systems ecology thus represented rival visions of how to construct a model of the ecosystem, the one based on the Darwinian principle of competition, the other on a more holistic vision of apparently purposeful feedback loops. Philosophically and politically, they invoked very different implications about nature and human society. The result was a deep level of conflict in which each side dismissed the other as philosophically naive and scientifically incompetent. The later twentieth century thus did not witness a unification of ecology around a coherent paradigm. There were still different schools with different research programs, methodologies, and philosophies. The one thing they all seemed to agree on was that scientific ecology had to present itself as essentially materialistic, offering no opening for communication with the kind of nature mysticism favored by the extreme environmentalist movement. Although systems ecology retained a holistic approach reminiscent of Clements's vision of the ecosystem as an organism in its own right, the advent of cybernetics and the link to economics allowed even this school to distance itself from the old idealism.

It is in this context that we can judge the reaction to James Lovelock's Gaia hypothesis of 1979, in which the whole earth is seen as a self-regulating system designed to maintain life. Gaia is the name of the ancient Greek earth goddess and was chosen to imply that the earth is mother to all living things, humans included. Lovelock made no secret of his support for environmentalism, criticizing those who advocate unrestricted exploitation of nature by implying that Gaia will, if necessary, take steps to eliminate humanity if it becomes a threat to the whole biosphere. Lovelock had impeccable scientific credentials, having worked in the space program developing systems to monitor the earth's surface from satellites, but the rhetoric with which he presented his theory clearly touched a raw nerve with many scientists. Although apparently similar to the systems approach, Gaia seemed to go beyond the cybernetic analogy and return to the older organicism in which ecosystems (in this case the biosphere as a whole) have a real existence and can act on their own behalf to achieve their own purposes. Critics were not slow to point out these implications, dismissing the whole theory as a perversion of science that pandered to the romanticism of the environmentalist movement. For Lovelock, it was as though a dogmatic scientific establishment had closed its ranks in defense of materialism: "I had a faint hope that Gaia might be denounced from the pulpit; instead I was asked to deliver a sermon on Gaia at the Cathedral of St. John the Divine in New York. By contrast Gaia was condemned by my peers and the journals, *Nature* and *Science,* would not publish papers on the subject. No satisfactory reasons for rejection were given; it was as if the establishment, like the theological establishment of Galileo's time, would no longer tolerate radical or eccentric notions" (Lovelock 1987, vii–viii). Nothing could more clearly indicate the gulf that still existed between scientific ecology (in all its forms) and radical environmentalism.

■ Conclusions

Although many people associate the term "ecology" with the environmentalist movement, we have seen that scientific ecology has a variety of origins, most of which were not linked to the defense of the natural environment. Science has more often been associated with efforts to exploit natural resources, and historical studies show that ecology emerged more from a desire to manage that process than to block it. At best, the majority of biologists have been concerned to ensure that humanity's engagement with the natural world does not do too much damage: sustainable yields are preferable to the wholesale destruction of a resource. Even those ecologists who imagined the ecosystem as a purposeful entity with a life of its own were willing to offer advice to farmers and others whose activities necessarily interfered with the untouched state of nature. In Europe, the whole idea of a purely natural landscape seemed meaningless, so ancient and so pervasive was the human role in shaping the environment. Although the more radical environmentalists can draw comfort from theories such as Lovelock's Gaia, they cannot lay claim to ecology as a science that inevitably lends support to their view that nature should be left untouched.

Equally interesting for the historian of science is the diversity of origins and theoretical perspectives from which the various branches of ecology emerged. Here was no single discipline shaped by a common research program and methodology. On the contrary, the movement toward what became known as ecology occurred in different places and at different times. The various locations of the scientists who became involved shaped the problems they sought to answer and hence the methodologies they thought appropriate. A technique that made sense on the open prairie of the American Midwest would have been inappropriate for the much-tilled landscape of Europe or the tundra of Hudson's Bay. Into these diverse environments came scientists with different backgrounds and interests; some were plant physiologists seeking to extend the experimental method to the interaction between plant and environment, some were biogeographers or taxonomists. All were driven by a determination to make the study of the interactions between organisms and their environment more scientific, but what they defined as "scientific" depended on their background and the problems they confronted. There was much suspicion to begin with over the application of mathematical techniques to modeling ecosystems. The majority of ecologists wanted to portray their science as materialistic, and this eventually led to link with the revived Darwinism of the evolutionary synthesis. But there has been a persistent current of philosophical opposition to this movement, paralleling similar doubts in other areas of biology. Smuts's holism was by no means uncharacteristic of a nonmaterialist current of thought in early twentieth-century science. It certainly appealed to some of the early ecologists, and although that way of thought became less fashionable in the late twentieth century, its revival in the form of the Gaia hypothesis ignited a new level of debate. This debate reminds us of the gulf that still exists between the majority of scientists and the almost mystical vision of nature that has sustained the more radical environmentalist movement.

References and Further Reading

1. Anker, Peder. 2001. *Imperical Ecology: Environmental Order in the British Empire, 1895–1945.* Cambridge, MA: Harvard University Press.
2. Bocking, Stephen. 1997. *Ecologists and Environmental Politics: A History of Contemporary Ecology.* New Haven, CT: Yale University Press.

3. Bowler, Peter J. 1992. *The Fontana/Norton History of the Environmental Sciences*. London: Fontana; New York: Norton. Norton ed. subsequently retitled *The Earth Encompassed*.

4. Bramwell, Anna. 1989. *Ecology in the Twentieth Century: A History*. New Haven, CT: Yale University Press.

5. Brockway, Lucille. 1979. *Science and Colonial Expansion: The Role of the British Royal Botanical Gardens*. New York: Academic Press.

6. Cittadino, Eugene. 1993. *Nature as the Laboratory: Darwinian Plant Ecology in the German Empire, 1880–1900*. Cambridge: Cambridge University Press.

7. Coleman, William. 1986. "'Evolution into Ecology?' The Strategy of Warming's Ecological Plant Geography." *Journal of the History of Biology* 19:181–96.

8. Collins, James P. 1986. "Evolutionary Ecology and the Use of Natural Selection in Ecological Theory." *Journal of the History of Biology* 19:257–88.

9. Crowcroft, Peter. 1991. *Elton's Ecologists: A History of the Bureau of Animal Population*. Chicago: University of Chicago Press.

10. Kingsland, Sharon E. 1985. *Modeling Nature: Episodes in the History of Population Ecology*. Chicago: University of Chicago Press.

11. Lovelock, James. 1987. *Gaia: A New Look at Life on Earth*. New ed. Oxford: Oxford University Press.

12. Leopold, Aldo. 1966. *A Sand County Almanac: With Other Essays on Conservation from Round River*. Reprint, New York: Oxford University Press.

13. Marsh, George Perkins. 1965. *Man and Nature*. Edited by David Lowenthal. Reprint, Cambridge, MA: Harvard University Press.

14. Mackay, David. 1985. *In the Wake of Cook: Exploration, Science and Empire, 1780–1801*. London: Croom Helm.

15. McCormick, John. 1989. *The Global Environment Movement: Reclaiming Paradise*. Bloomington: Indiana University Press; London: Belhaven.

16. Merchant, Carolyn. 1980. *The Death of Nature: Women, Ecology and the Scientific Revolution*. London: Wildwood House.

17. Mitman, Greg. 1992. *The State of Nature: Ecology, Community, and American Social Thought, 1900–1950*. Chicago: University of Chicago Press.

18. Palladino, Paolo. 1991. "Defining Ecology: Ecological Theories, Mathematical Models, and Applied Biology in the 1960s and 1970s." *Journal of the History of Biology* 24:223–43.

19. Sheal, John. 1976. *Nature in Trust: The History of Nature Conservancy in Britain*. Glasgow: Blackie.

20. _____. 1987. *Seventy-five Years in Ecology: The British Ecological Society*. Oxford: Blackwell.

21. Taylor, Peter J. 1988. "Technocratic Optimism, H. T. Odum, and the Partial Transformation of Ecological Metaphor after World War II." *Journal of the History of Biology* 21:213–44.

22. Tobey, Ronald C. 1981. *Saving the Prairies: The Life Cycle of the Founding School of American Plant Ecology*. Berkeley: University of California Press.

23. Worster, Donald. 1985. *Nature's Economy: A History of Ecological Ideas*. Reprint, Cambridge: Cambridge University Press.